QUIRK

QUIRK

BRAIN SCIENCE MAKES SENSE
OF YOUR PECULIAR PERSONALITY

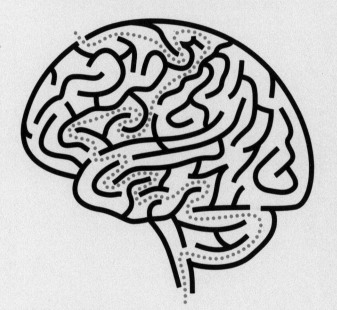

HANNAH HOLMES

RANDOM HOUSE NEW YORK

Published in the United States by Random House, an imprint
of The Random House Publishing Group, a division of
Random House, Inc., New York.

RANDOM HOUSE and colophon are registered
trademarks of Random House, Inc.

ISBN 978-1-4000-6840-1
eBook ISBN 978-0-679-60452-5

Printed in the United States of America on acid-free paper

www.atrandom.com

2   4   6   8   9   7   5   3   1

First Edition

Book design by Christopher M. Zucker

FOR JOHN

CONTENTS

FACTOR: OPENNESS

INTRODUCTION

"EACH OF US IS SPECIAL."

"Each of us is unique."

"And every human embodies a magical upgrade, a soul, a spark of the divine, which elevates us above all other creatures."

Spark, schmark! That's my private title for this book. Humans have no more sacred spark in our personality than squirrels do.

I reached this conclusion when researching depression. Researchers studying the disease were working with a mouse they had engineered to be depressed. If mice have enough personality to mimic our depression, I thought, what else do they have going on?

Quite a lot. Almost everything. Even an artistic side, and something akin to religion. In those tiny heads a full-blown personality guides each animal's behavior in a slightly different way. Mice have a whole lot going on. Or, to reverse the animals: Humans don't have much more going on than mice do. In the fundamentals of personality, we can pinch-hit for each other.

How did this come to be? It happened because life on earth is chancy. In the pursuit of successful reproduction, every animal must navigate the equivalent of cats trying to eat you, weasels trying to cheat you, and a flood carrying away your winter's supply of food. Life is risky. And the key to personality is that there's no single solution that answers every risk.

For instance, how best to deal with cats trying to eat you? This spectrum of personality is summed up as approach/avoidance. The approach-oriented mouse is calibrated to shrug off risks and charge into the world seeking new foods, meeting many mice, exploring new basements. The avoidance-oriented mouse is tuned to stay home and avoid exposure to cats, which also limits her exposure to fresh opportunities.

Cheating weasels present the quandary of how much a mouse should invest in relationships. Allies support you when you're in trouble, but they also demand a percentage of your time and energy—energy you might spend on your own precious self. The social mouse invests in winning the weasel's loyalty. The independent mouse cheats that weasel right back and takes her chances going it alone.

Flooding is even less predictable than friends. In an unstable world, is it best to live fast and die young? Or should you keep your nose to the grindstone, stocking away resources for the future? The short-term mouse skimps on the food-storing and the friend-grooming, and spends more time on eating, breeding, and otherwise seizing the moment. The long-term mouse puts in the effort up front, so that when the environment does falter, he has a backup plan.

And so it is with us humans. Spark, schmark. We're all just trying to beat the system. And again, there is no right answer. Well, actually, any personality that works—that gets an animal across the finish line with healthy offspring—is a right answer. That personality, carried in the genes of the offspring, will win the right to go another round.

It's liberating. I am far more at ease among my fellow mice than I was before writing this book. Once, I took others' personalities personally. When someone was rude or selfish or irresponsible, I winced.

But personality isn't personal. It's biological. It's a series of dials— Extraversion, Neuroticism, Agreeableness—each set to a different temperature. Even the "nurture" half of personality, which we once thought could counteract the nature of genes, now looks biological. I was a gregarious, if shy, child born to two loner parents. But rather than passively allowing my environment to quash my social nature, my genes prodded me to seek out people, people, and more people. In school I joined teams and clubs. In adulthood I fled the rural life for a perch in the city, among flocks and flocks of people. A personality creates the environmental conditions that it evolved to thrive in.

Much of what we're learning about human personality we owe to the mice. Of course, scientists don't use animals for spurious inquiries such as why people prefer vanilla or chocolate. They use mice to understand diseases of personality—depression, anxiety, addiction, psychosis. It's through the lens of how a brain goes wrong that we're learning how a normal brain functions. And it's through the mice that we're finding better drugs, and environmental treatments, too, for those brains that malfunction. Just as mice have brought us treatments for cancer, they're now providing relief from mental disease.

It may seem ludicrous that the basics of personality can be written in a brain so small. But again, all animals evolved with the same pressures bearing down upon them: Go forth and eat! Go forth and multiply! Hang back and don't get eaten! Join the group and share the burden! Protect your turf and share only with your own offspring! And those pressures forged the universal dimensions of personality, the dials that come preset in every mouse and man.

Two things distinguish the human personality from that of a mouse. One is our profoundly social lifestyle. Most mammals evolved to fend only for themselves, but a few species found that the benefits of cooperation outweigh (if only by an ounce) the self-centered simplicity of a solitary existence. Our social life is etched into the personality of our entire species. Instinctively, we communicate. Biologically, we're built to share. Without ever meaning to, we care. Not everyone cares equally, but even the nastiest person you know cares more than the nicest weasel or bear.

Our other distinction is the sheer size of our brain. Our tremendous wattage, plus the social instincts, yield nuances of behavior that we don't see in other creatures. And when the nuances mingle and collide, amplifying or offsetting one another, our personality becomes complex. Not magical or divine, but really, really complex. Quirky.

So perhaps every personality on earth is, in fact, unique. Not because we're each a spark of some divine entity. Evolution rewards diversity; lately science is discovering this is true in many natural systems. And personality appears to be just another of those.

THE FIVE FACTOR MODEL

THIS BOOK BORROWS THE Five Factor Model to make sense of your personality. The model slices your temperament into five major factors, then divides each factor into six facets. Herein you'll find that same organization—for the most part. Personality is insanely hard to summarize, whether you use five factors or eighteen thousand.

Perhaps that contributes to our timeless fascination. But humans are also deeply social creatures, and our day-to-day success is deeply dependent on working with the people around us. We invest heavily in studying one another, and ourselves, because with understanding comes mastery. So we try, and try again, to find the hidden system that would explain our personal peculiarities.

Are you phlegmatic or sanguine? An Aries or a Libra? An ENTP or an ISFJ? An ox, a rat, or a rooster? Are you irascible, irreverent, or irritating? Appeasing, approachable, affable? What are you? *Who are you?*

We're horribly complicated, that's what we are. Just when you think you know someone, she does something unexpected. Wouldn't it be nice if there were a simple system, a reliable test, that could describe people accurately?

The search for such a system or test has been going on for a long time. The Western zodiac goes back almost 3,000 years; the Greek bile-blood-phlegm scheme predates the Myers-Briggs scheme by at least 2,400 years. And the sheer number of words we use to describe temperament hints at how much personality matters.

In fact, a blizzard of words marked the birth of the Five Factor Model. The process began with Sir Francis Galton, a cousin of Charles Darwin, who invented nearly everything. To Galton we owe fingerprinting, meteorology, a hearing test, the personality questionnaire, the concept of nature vs. nurture, and the "lexical hypothesis." He held that when a feature of personality becomes critical to how we deal with one another, that feature will find itself a word. Hence "feisty," "ferocious," "fickle," and "flippant."

True? That's still open to debate. But the idea inspired a couple of guys to excavate 17,953 personality words from the dictionary in 1936. By eliminating redundancies (kind/compassionate, churlish/cross) they reduced the human personality to 4,000 words. That's still a bit unwieldy, if you're looking for a simple system, a reliable test.

In the following decade another guy compressed the list to 171, then managed to divide those into thirty-five categories, which he finally squeezed into sixteen overarching factors. But it was the United States Air Force that bombed human personality into submission. In 1961 a couple of military psychologists compacted us into just five factors, acronymed as CANOE or OCEAN: Openness, Conscientiousness, Extraversion, Agreeableness, Neuroticism. Each factor splits into six facets, for a total of thirty dimensions. The model has been translated for non-English cultures, and seems to hold up well.

But variations abound, and bickering continues.

For instance, the Five Factor Model leaves out some people's pet features. Why is humor not included among the thirty facets? Or grumpiness, gullibility, argumentativeness, absentmindedness, or sensuality?

And it's not very, well, scientific. It is, at its core, a list of terms, wrung out of the 18,000 words that English speakers use to describe one another. It describes personality, but does nothing to *explain* personality. A facet—impulsiveness—is a paper label pasted on a gear in a machine, but it doesn't tell us anything about the gear's function.

That said, the Five Factor Model, a.k.a. the Big Five, has struck a balance between simplicity and richness that most psychologists can live with. It's not perfect, but it's the best we've got. Today it has become the workhorse of personality research. It allows scientists to estimate the influence of genes on a personality trait. More immediately, it helps psychologists to identify people at risk for personality disorders such as depression, anxiety, and drug addiction. Therapists employ it as a guide to a client's natural inclinations. It's even challenging Myers-Briggs as the go-to test for building a healthy and harmonious workplace. The applications in daily life are truly boundless, judging from research paper titles like this: "The Big Five and Marching Music Injuries."

Even the people who study the flesh-and-blood machinery of personality agree it's a tolerable model. The psychiatrists and neurologists who research personality by looking at our genes, nerve cells, and brain chemistry appreciate that a person's score on the Five Factor Model can sometimes predict what they'll find under the skull. A person whose Five Factor score indicates high anxiety may well have a more active amygdala deep inside her brain, for instance. A person whose impulsivity score is high also has higher odds of sporting a lackadaisical prefrontal cortex.

And sometimes it can't predict anything. When I set out to find the biology behind those thirty facets, I found that sometimes a facet proves to be just a word hanging in the air. Sometimes the gear it names appears to have no home in the brain. Anger, for instance, is a facet of Neuroticism. But anger lights up no distinct regions of the human brain, nor is it a trait we can identify in mice. Other facets look like redundancies, where two names have been pasted on to the same biological gear: Is there much difference between "impulsive" and "immoderate"? Brain scientists have reached the same conclusion. When they compare our Five Factor personality to our brain, they simply ignore many of the thirty facets.

So the model is not a foolproof measure of personality. Increasingly, brain science reveals the weak points. But what measure of personality is foolproof? Even in its frail state, the Five Factor Model is a lot more detailed than the Myers-Briggs, less starry-eyed than the zodiac, and much less funny than the theory of the humors.

This book follows the Five Factor Model, but not blindly. When I encounter a Five Factor label that gets no support from flesh-and-blood science, I am going to cut that word loose, sending it back to the pile of 18,000.

Facets that hard science can support will follow this format:

A mini-questionnaire introduces the facet. Your answers will give you a general sense of where you stand on that element of personality. For a more thorough analysis of yourself, complete the full questionnaire at www.hannahholmes.net.

Next we'll meet a mouse strain (or in some cases a rat or vole strain) created to embody that aspect of personality.

Then we'll see how research into human personality adds to what we've learned from the mice.

Finally we'll drop into deep time to see why evolution might have allowed that facet to persist in such a wide range—why, for instance, some mice are rowdy while others are retiring.

1
FACTOR:
NEUROTICISM

NEUROTICISM FACETS
Anxiety
Depression

THE CAST

Mascot . . . . .5HTT KO;
also introducing Mitzi and Maxi

Neurotransmitter . . . . .Serotonin

Brain Regions . . . . .Amygdala, Prefrontal Cortex

NEUROTICISM

ARE YOU THAT PARTYGOER who neglects the stuffed mushrooms because you're a little freaked out by that guy scowling alone on the sofa? Or perhaps you avoid the mushrooms because you're worried that other people will see you pigging out. Perhaps you just aren't hungry, the party is a disappointment, and you wish you were at home in bed. Maybe you're the partygoer who does eat the stuffed mushrooms, about half of them, because once you start you just can't stop. If any of these sound like you, congratulations! You could have a Neurotic personality!

Neuroticism is about avoidance. Neuroticism is about anticipating the worst and retreating. It's about being the first person to come in out of the rain, and the last to believe the storm has passed. A Neurotic brain is attracted to angry faces, to shouts of alarm, to catastrophic headlines. It commits life's harshest lessons to memory and does not forget them.

For a strongly Neurotic person life is not a bowl of cherries. A Neurotic brain is more upset by stressful events than a less Neurotic brain would be. A person with a Neurotic personality is inclined to be moody and emotional, at best. At worst Neurotic people have a higher chance of sliding into a full-blown anxiety disorder or depression. Neuroticism is a bit of a burden.

But somebody has to be the doubter. We can't all just laugh and play the days away. Someone has to look gift horses in the mouth. Someone has to interrogate the charming stranger before we throw open the gates. Someone has to look past the sunbeams and notice the gathering thunderheads. People with Neurotic brains are our watchdogs. They notice signs of trouble first, they remember every disaster that ever happened, and they're always on guard against new ones.

Serotonin is the signature brain chemical of Neuroticism. This chemical helps information move through our (and every animal's) nervous system. If you have the normal amount of serotonin lubricating your brain, you've got a shot at serenity. But if your serotonin is either too low or too high, Neuroticism may cast its shadow over your picnic.

The amygdala is the mascot brain region of the Neurotic personality. Named after an almond, and tucked deep between your ears, it's an ancient emergency control center. This little nut has the power to dismiss your fancy intellect when circumstances call for action instead of analysis. When the amygdala detects an emergency it prepares you to run, fight, scream, or overturn a vehicle in order to protect life and limb. In a person with a Neurotic personality the amygdala is set on "supersensitive." It's jumpy. It's trigger-happy. It's an emergency center that would rather issue a false alarm than be caught off guard. In fact, the more stress you throw at a Neurotic amygdala, the more watchful it becomes.

But Neuroticism does more than keep us all safe from lightning and spiders and mayonnaise left out overnight. A recent study found that anxious children are less likely to die in stupid accidents—the "Hey, watch this" accidents that claim carefree kids who ride skateboards down stairs or play with loaded guns. The Neurotic personality also appears to convey a certain blessing of intelligence, particularly the planning variety that helps a person to avoid surprises.

The anxiety and depression facets of Neuroticism are heavyweights in human happiness. If you score relatively high for those, that doesn't mean you're destined for the psychiatric ward. But a personality that's high in anxiety or depression does have an elevated risk of developing mental illness. These personalities are poised to skid right off the Neuroticism personality scale and plunge into the pit of psychiatric disorders. Anxiety disorders and serious depression are painful conditions. So science hunts for a cure. And in the process, science reveals a lot about personality, both disordered and normal.

THIS GIVES YOU A QUICK look at where you land on this facet. If your answers tend toward the "often" side, you're higher in that facet.

Anxiety is the quintessential "avoid" emotion. When it rises up, it's instructing you to step away from the edge of the cliff, back away from the spider, run away from the man with the axe. Everyone's brain monitors her environment for danger. Some brains shrug off most of the omens and portents as meaningless, and others duck and cover for every passing sparrow.

Anxious Mouse

The father of serotonin, Klaus-Peter Lesch, resides in Germany. On a bleak November day I travel to Bavaria to make his acquaintance. And to meet the mouse version of myself that he has created. My train rushes into a hilly eastern quadrant of Germany I've never seen before. The leaves have fallen, but ivy and other creeping things keep the woods green, and under spitting clouds the farmland looks mossy. Then we wend into a river valley with steep golden hills on either side, and I've arrived in Würzburg, home to the anxious mouse.

Wait, steep golden hills? What is that gold stuff? Grapevines? Seriously? Just my luck! Normally I am a traveler whose road stories detour into descriptions of the ceviche, or the curry, or the smoldering chestnut liqueur. Approaching Germany I steeled myself for the würsts of many colors, but I neglected to review the beverage category beyond a quick nose-wrinkle regarding beer. Now here I stood in the cloying, the unbearably sweet, heart of Riesling country. Riesling is the wine I love to hate. This far north, a grape's growth is retarded by the low temperature, allowing sugar and acid to accumulate under its skin. If the wine's fermentation ceases before that sugar is converted to booze, then achtung, baby! Sticky wine warning! Oh, well. This is work, not play.

In the morning I walk across town to find Lesch, who works in a building complex among the vineyards. "Nervenkliniken," the sign on the guardhouse reads. This I find charming. Technically, it means "neural clinics." But it reminds me of the olden-days term for people like me: I'm "nervous." I've got "weak nerves." I must "take the cure at a *Nervenklinik*." It occurs to me that the standard olden-days medicine for a case of nerves—alcohol—was squeezed from the grapes now being displaced by the *Nervenkliniken*. Times change, cures change.

My goodness, it is a nice *Nervenklinik*. It's airy, and tropical plants crowd every window. And let me just say two things for the record, two things that may be applied to every German academic office I enter in the course of this book. First, they're so nice. The paint is fresh, the carpeting—carpeting!—stands proud, the furniture is neither rusty nor peeling. Second, they're tidy. I've been studying scientists for twenty years, and I've never seen so many bare desks! A more typical academic habitat contains so many research papers and journals and books that the floor (scuffed linoleum, usually) is visible only in little trails the resident treads between the piles. Here in Germany an office will have a wall of closed bookcases and file drawers, leaving the entire floor and desk surfaces exposed! I swear I'm not falling for the cliché regarding German orderliness. It is a truly striking difference in the culture of academic shelter.

The tenant of the proverbial corner office is a towering specimen with graying hair a few millimeters longer than a crew cut, and a wide, tight smile. Among the requisite plants on the windowsill are photographs of his blond family. Lesch is a psychiatrist with one hiking boot in the human world and the other in mousedom. He ministers to human patients who battle their depression in an adjoining hospital. And in a nearby laboratory, he alters the genes of mice to create animals prone to depression. By studying the mice, he hopes to find a cure he can transfer to those humans in the hospital.

"Coffee?" he asks, hoisting his long frame away from the computer and toward the coffee table. "No? No coffee? We drink a lot of coffee around here!"

With my first question, he lurches back to the computer to print off a research paper. Then returns. On the occasion of my third question he retrieves a laptop and starts flipping through PowerPoint presentations. "It's easier to describe if I have a few slides [tap tap] . . . a talk I just gave in London . . ."

But here's the short story: Back in 1996 Lesch discovered what one colleague has described as a needle in a haystack—he found a serotonin gene that caused measurable differences in personality.

He wasn't searching randomly. Scientists already knew serotonin probably had a role in anxiety and depression. And they knew where to find some of the genes that operate the serotonin system.

Serotonin is an ancient chemical, found in every living thing endowed with a nervous system. (It shows up in plants, too, but as far as we know does not make them moody when it malfunctions.) The stuff is made in special cells originating in that gnarled old tree trunk of the brain, the brain stem. Most serotonin neurons stretch their axon arms down the spinal cord, to service the gut and other humble functions. But a few meander up into the brain, twisting and branching like morning glory vines. And like a morning glory vine, each axon is studded all over with buds, called spines, from which serotonin ultimately flows to work its magic.

Most simply, a serotonin molecule is a key. Its job is to linger in the intersection between two nerve cells, ready to help messages cross. When a message arrives, serotonin opens a locked gate on the next nerve cell so that messages can travel smoothly through the brain. Imagine the synapse as an intersection where three streets meet, each blocked by a gate. Two of the streets are axons that help move your arms. The third street is the spine of a serotonin cell. Way out on the front of your head, your nose itches. The neurons of your brain start to pass a message, fire brigade–style, with the goal of raising your hand to your nose. When the message reaches our three-way intersection, it pauses for a split second. It can jump across only if serotonin has already opened a gate across the way. So if the serotonin isn't flowing at the right speed…the message… stops…moving. Brain…must find a work-around. Start a new…message. Things move slowly. You, from the outside, could almost look… depressed.

By the time Lesch was groping for his needle in a haystack, science already knew serotonin was implicated in depression and anxiety. In the 1950s scientists noticed that a family of heart and lung drugs was making patients unexpectedly cheerful. Initially they dismissed this as a side effect: Who wouldn't be happy to recover from tuberculosis or heart disease? But in a human, an increase in happiness is a transient phenomenon. It lasts a day, a week, maybe a month, then the human reverts to the old baseline. These patients were staying happy much longer than they should. The best guess was that the drugs were increasing the amount of serotonin in their synapses.

Since then we've learned that this chemical has a central role in human personality. And monkey personality. And, yes, mouse personality.

Lesch's big breakthrough came in the form of a gene—or rather, two variations of the same gene, SLC6A4. This gene helps to manufacture a street-sweeper molecule that collects serotonin from the intersections in your brain. The serotonin neuron steadily dumps serotonin into the street; the street-sweeping molecule, or transporter, steadily collects it and returns it to the neuron for reuse. Lesch's street-sweeper gene contains a paragraph of DNA dictating how hard the gene itself shall work at making street sweepers. The long version of the paragraph makes a lot; the short version of the paragraph makes fewer. Each of us gets two copies of this gene from our parents, for a total of three combinations: long-long; long-short; short-short.

Lesch's discovery was that those people who inherited the short version of this gene from one or both of their parents ranked higher than normal on tests of Neuroticism. It was a reassuring confirmation that a balanced serotonin system is crucial to your happiness.

For me, reading Lesch's study was more confusing than reassuring. I had understood that we humans who suffer from depression and anxiety don't have enough serotonin sloshing around in our synapses. We take drugs that actually sabotage the street sweepers so that serotonin lingers longer in the intersections. Now Lesch is saying just the opposite: Too much serotonin in the synapse can also spoil your mood.

Researching on my own, I had furrowed my brow at this paradox until my brows ached. In Lesch's office, I aired my grievance: A lot of research says that too little serotonin messes you up. But other research says too much messes you up. So which is it?

"Both," he said, looking a little aggrieved himself. He's standing now, as though he is physically unable to be still. My father used to pace like this. I wonder if Lesch is a bit anxious. "You want just the right amount, and you want it in the right place," he says with a grimace. He thinks maybe an imbalance between the serotonin that's inside a neuron and out in the synapse—just that imbalance—creates an anxious personality.

When I got past my confusion, I realized this wouldn't be a novel arrangement. You could say the same thing for salt. If you don't have enough salt in your body, your nervous system sputters and stalls. But too much salt sucks water out of your cells, and then chews up your kidneys on the way out of your body. There's a happy medium where your nerves hum and your kidneys whistle and all is right with the world. And ditto for thyroid hormone: Too little and your metabolism burns too slowly, leaving you chilly, weary, and teary. Too much and you become hot, jittery, and hungry.

And so it seems with serotonin. In the correct amount it produces a calm animal who can tolerate her neighbors. When it's out of balance, both mouse and man are at risk for negative emotions: anxiety, depression, aggression, obsession, or all of the above.

In a stressful environment like our stimulating culture, a small imbalance in your serotonin can grow into a big problem. The Neurotic personality can easily become a clinically depressed personality, or a socially anxious personality. And then the serotonin-modulating drugs can become a multi-billion-dollar industry.

That's where the mice come in. You can't go around popping experimental pills into depressed and anxious humans. You can't take samples of human brain tissue to see if your pills altered serotonin distribution. I don't know about you, but even when I'm anxious, I still prefer to have my brain nearby. So the early testing of drugs rests entirely on the backs of mice.

Since his pivotal discovery Lesch has gone on to create mouse strains whose serotonin system is altered in a variety of ways. When he disabled the street-sweeper gene completely, he made a mouse who behaves, well, like me. Like someone who looks gift horses in the mouth, and peers through the sunbeams in search of thunderheads.

Lesch isn't the kind of guy who can abandon his multi-million-dollar research empire to make introductions between anxious humans and anxious mice. But he is the kind of guy who's surrounded by two dozen fellow geniuses. A postdoctoral student, Thomas Wultsch, draws the short straw. To visit the mice he leads me on a path through the vineyards. Crows turn like black leaves against the clouds. A student on a bicycle whispers past on damp pavement. Wultsch excuses his boss's absence.

"If I had to follow in his steps for one month I would fall down," he marvels. "He works very, very hard. And what does he do on his vacations? Climbs mountains. Nothing under four thousand meters." (Math moment: 4,000 meters is 13,000 feet.) Wultsch himself has thick black hair, heavy glasses, and a substantial burden of shyness. He's a model scientist, as long as you understand that many young scientists are quite hip. He's sporting cool shoes and an iPhone.

As we cross a parking lot, he glances down, and says, "Ten meters below is a home for forty thousand animals." Oh. That explains the stairways rising out of the cement—emergency exits.

Getting in is a lot harder than getting out, however. Until this day I had no idea what a threat I pose to rodents. I'll be allowed in only the dirtiest of three separate mouse stables, but still I prepare like a surgeon. Shoes off, clothes off, jewelry off, surgical scrubs on, then lab coat, then wash hands a long time. Now as I cross a line on the floor, I slip feet into rubber clogs. Enter prelab airlock. Hang up lab coat, step out of blue clogs. Step across another line on the floor. Zip into white fiber zoot suit. Add green hair net. Rub hands with alcohol gel. Pull on latex gloves and face mask. Step into green rubber clogs.

"Can I bring my notebook and pen?" I plead.

Wultsch levels his serious glasses at me. "Has it been in an animal-breeding facility in the past forty-eight hours?"

"Not that I recall, but I slept for part of the flight."

"Do you need to use the washroom?" he asks somewhat belatedly. "Once you're in here, you have to do this all over again to go out. Sometimes when we are doing an experiment we have to do it all at once, so we are in here for ten or twelve hours with no food, no water." We step over the last line. I want to meet my mouse.

This is the "clean lab," an immense condo development for the mice used in experiments. A row of doors on either side of a hallway reveal rooms where stainless steel racks hold clear plastic mouse houses. These are not cheesy condos. This is about as luxe as a mouse's life can get, except for a few minutes on experiment days. And even then, I bet the wild mice in my cold, damp cellar would trade places in the twitch of a whisker.

The heavy doors protect the mice from random noises that might interrupt their slumbers—mice are nocturnal and sleep during the human workday. A digital device in every room tracks temperature and humidity—mice prefer a relative humidity of 50 percent. Their bedding is of sawdust so fine and soft that it goes airborne when the mice scamper. This is a hassle for humans, but as the mouse vet asked me later, "Do you prefer to walk on a sandy beach, or a stony beach?" Yes, the mice have their own vet, and they never get sick, because germy people like me are wrapped up like bad meat. The mice also get a fresh paper towel (unbleached) each week when their condos are cleaned. The towel provides hours of entertainment as they reduce it to preternaturally uniform shreds and knit it into a fluffy igloo. Their days they spend snuggling inside the igloo, says Wultsch, and their nights they spend remodeling. Day and night are dictated by a timer. If humans need to work with the mice during the mice's active time, it's the humans who alter their sleep schedule, not the mice. These are very coddled mice.

Wultsch uses the last room on the hall to run experiments because no one ever walks past that door, potentially distracting his animals. Only a handful of people have keys to start with—the vet, a few researchers, and the technicians who care for the mice. But by positioning the behavior lab at the end of the hall, Wultsch avoids even the small chance of distraction.

"We have to make the mice feel as comfortable as possible, and we have to get as familiar with them as we can," Wultsch says, opening this final door. "Normally when I come in the morning I start talking to them right away: 'Good morning, mice. How did you sleep?' "

The second thing I notice is the sound: scuffling and scuttling and skittering of mouse bodies. But the first thing I notice is the adorableness.

These are a variation on the "Black 6" mouse, a common research strain with dark, glossy fur. They have shiny eyes and translucent, smoky ears. Whiskers as fine as spiderwebs fan from their noses. Despite the time—it's a few hours past bedtime, for a nocturnal critter—they're all up and about. They're rearing to gnaw at pellets in a feeder, they're stretching toward the ceiling to sample the air, they're working the water bottle, they're sniffing each other front and back like dogs. The skritching of paws in shavings is constant.

I bend to peer into a condo. A black face glances my way, but then rears for the ceiling again, sampling the smells that ooze from my zoot suit. Mice rely very little on their eyes, and can see approximately a foot. But in the next twenty-four hours I'm going to see how quickly they sum up a human using other senses. "The animal first identifies you by voice, then by smell," Wultsch says. "I ask new students, 'Please, do not change your shampoo in the months that you are here. If you one day wear your new perfume that you got for Christmas, I can tell you your experiment will go wrong.'" According to a mouse's sensory system, a new smell amounts to a mysterious stranger in the room.

Wultsch selects a box with three mice in it, and sets it on a bench. "These are for you." Three mice, engineered to be anxious. Let's see how they run. Let's see if they approach, or avoid.

The benefit of a genetically anxious mouse is that researchers can test drugs on it. But how do you know if a mouse is, or is not, a worrier? Looks can be deceiving, in both mice and men. Most of my friends are surprised to learn that I am fundamentally anxious. It's the same with these mice. They're eating and sniffing and cuddling like any other mice I've known.

Behind a black curtain, in a black booth, stands one of the classic devices for the measurement of mouse angst. The black booth minimizes distractions and puts the mouse at ease. But the elevated plus maze does not.

It's not the sort of maze a mouse could get lost in. It's just four arms radiating from a central square, forming a plus sign. But two of the arms are made cozy by high walls. The other two are open—open, in the mind of a mouse, to owls attacking from above, and foxes attacking from every other direction. The apparatus is elevated to dissuade mice from hopping off and pursuing more gratifying pastimes. The height, which seems infinite to a half-blind mouse, makes those open arms even more "aversive," as scientists say. Especially to an anxious mouse.

Lifting the lid off the condo, Wultsch coaches me. "When you pick her up you must know which one you will get, and don't hesitate," he instructs. "If you are anxious, they will become anxious." Or even more anxious, as the case may be. With their ceiling gone, the sleek youngsters stretch up the walls, sifting, sifting.

I track one mouse around the condo, then swoop, and lift her by the tail. She wriggles a bit, but she's used to such flights. I'd love to cup her in my hand and pat her, but that's bad form. I elbow aside the black curtain. My flying mouse must land in the center of the maze, facing an open arm. I lower her smoothly until her front paws touch, then her hind paws, then I back out and leave her to face her demons.

By the time I duck back into the foyer, Wultsch already has her on the computer monitor. A dim light over the maze allows a camera to capture her movements. So there's my mouse, noodling out along the scary, open arm I pointed her toward. Her nose is up, she's whiskering the air and embracing the explorer within. But about halfway out, she undergoes what a psychologist might call a triggering event. This happens all the time to humans: If you are born with the genetic capacity for anxiety, a triggering event can hit the switch and bring your worrying into full bloom. For my mouse, the triggering event arrives in the slip of a paw. Her left hind leg slides over the tiny lip on the arm; then the right leg follows. She fights to pull herself up, the hero in her own action movie. When she does heave back onto the arm, she lowers her head and scoots for the darkness of the closed arms. Even if she wasn't feeling anxious before, she will be now.

Once in shelter, she regains some confidence. Again with nose extended, she whiskers down the left wall of a closed arm. She turns at the end and whiskers up the right wall. But when she reaches the exposed crossroad, she halts. She peeks out at the arm on her left—open. She takes one exposed step, then a second. And that's enough. She retreats. Then she scoots across the intersection to the other closed arm. As before, she passes down one side, then up the other. Again when she reaches the crossroad, she stops. She peers at an open arm but does not go. Her inner explorer has been tied up in ropes of anxiety. She doesn't venture down an open arm again. At the end of five minutes, I reenter and gently corner her so I can catch her tail. Back in the condo, her two friends sniff her face, and the place where my gloved hand has gripped her tail. Gradually all three return to their mousy duties—nibbling kibbles, digging in the sawdust, shredding the paper towel.

Wultsch's computer has recorded her track, and a ratio of time spent in open and closed arms. She is typical of this strain of Black 6, whose genes tilt them toward greater fear. These mice average about nine seconds in the open arms, while a normal Black 6 mouse might explore for four times that long.

"It gets boring after about ten mice," Wultsch admits, clicking the video shut. "I don't think it would be a success on YouTube." But human eyes must monitor every mouse in the maze. The mouse version of panic, which is freezing, usually strikes on the open arms. As a result, the mouse spends more minutes in the open than in the closed arms. Although a frozen mouse is motionless, the computer doesn't account for that. When it tallies the minutes spent on the open arms, it will assign such a mouse a ratio that describes a bold animal who hardly bothered with the sheltered arms. Behavior, whether of mouse or man, is not always what it seems.

Here, in the 5HTT KO mouse, is an anxious animal. Lesch has "knocked out," or silenced, the gene that builds her serotonin street sweepers. In the intersections of her brain serotonin molecules are piling up. And too much of a good thing is a bad thing. Lesch has measured the overload—she has about eight times the normal amount of serotonin in her synapses. But inside her serotonin-making cells there is nearly none. Her brain is out of balance. It makes her fearful. And it makes her resemble me.

Wultsch offers to let me run the other two mice but I don't want to waste them. My unfamiliar voice and odor would ruin the data they produce. And for some experiments, a mouse can be used only once. The second time you put a mouse into a maze, you lose the element of surprise. The mouse is familiar with the situation, and will behave quite differently. Remember the first time you went to a dentist, having no idea what would be done to you? That chair! Those steel drills! But the next time you went it was either more or less alarming, depending on how the first visit went.

Mice are the same. That doesn't mean they're useless after one trial. They can go on to tests of their intelligence, their social skills, their memory, and so on. Or, they can be treated with a potential anxiety drug, and retested on the elevated plus maze, to see if there's a change.

Furthermore, these black beauties aren't cheap. I wouldn't ask Lesch to squander his genius calculating the cost to produce one anxious mouse, but it's a pretty penny. With your average gene-altered mouse, it can cost $100,000 to create the strain, even after you've zeroed in on the gene you want to alter. Add the cost of room and board for a herd of them ($144 a year, per mouse, at one U.S. university). Even to buy very common strains of mouse off the rack, so to speak, runs $15 to $100 per head, and that's from a nonprofit mouse bank. I wasn't playing around with a million-dollar racehorse, but neither was I holding a two-dollar "feeder mouse" like Mitzi and Maxi, who live on my desk at home.

We bid the mice good day and let them return to their beds. Wultsch introduces me to some of the other residents of the mouse lab. It will take more than one kind of mouse to solve the riddle of anxiety. Humans demonstrate many different kinds of anxiety, and apparently the same is true for mice. Scientists keep creating new strains and learning new ways that anxiety takes root in a brain.

The STIM2 KO is an anxious mouse that has no trouble solving a traditional maze—until it undergoes a stressful event. After being stressed by a mild shock this mouse can learn nothing. It becomes a mouse moron. Another mouse, nicknamed Kevin, begins life with an anxiety score even lower than a normal mouse. But one stressful event is all it takes to transform Kevin into a hyperanxious mouse who responds to the world with great trepidation. Around the world, dozens of other strains are helping to solve the anxiety puzzle.

The altered genes in these various mice effect not only serotonin but other signal-sending chemicals as well. Some of these mice have normal serotonin systems. And one, also here in Lesch's stable, has no serotonin in its brain at all. You might expect this mouse to be a basket case, chewing its paws and lying awake all day worrying about cats and rats. Not this mouse. This mouse is physically and mentally healthy.

There's more than one way to make an anxious mouse. And the same goes for us. I suppose it's even possible that a few humans walk among us with no serotonin in their heads, happy as can be.

Anxious Human

It's summer, I'm ten years old, and I'm walking the gravel shoulder of a country road to my cousin's house. I hum a tune to myself and kick pebbles in the sunny silence. But inevitably the silence gives way. A couple of hills away an approaching car telegraphs its intentions. As the sound grows, now just one hill away, I steel myself. No longer humming, I force one foot in front of the other. But my eyes now skitter to the roadside ditch, and the forest beyond. The motor hum swells into a rumble and now I see it. In half a heartbeat I'm down in that ditch. I am flat in that ditch with my cheek pressed to the weeds and the cigarette butts. I am flat in that ditch and I am not breathing.

zzzzzZZZZZZZZZZOOOOOOOOMMMMMMmmmmmmmmmm

The car does not stop and I'm not pursued into the woods and slain. The sound diminishes with the next rise, fades on the descent. I inhale. A few seconds later I climb to the shoulder and resume my pilgrimage.

I didn't do it for every car. Sometimes I could control my legs, willing them forward step after step. But with each approaching vehicle the need to hide rose up until I nearly froze. In retrospect, I was lucky I lived on a little-traveled road. If I had grown up on a thoroughfare it would have taken me days to walk those two miles. Heaven knows what my cousin thought when I showed up with gum wrappers in my hair. She had issues of her own. Blind to the menace of passing cars, she would stop and orchestrate a somber burial for every dead bird and mouse she encountered.

Yeah, I was anxious from an early age. Now that I think about it, there was an earlier event that signaled a slight overtightness in the nerves. I grew up on a small farm in rural Maine, miles and miles from a streetlight. Crossing the lawn one dark night (eek!) to shut the chickens in their coop, I heard what I understood to be a rabid fox. They were around that year, or talk of them was. So, I heard one bark. I sprinted back into the barn, cut left into the feed room, clambered into a wooden barrel, and again ceased breathing. There in the darkest dark I crouched, marking each second with a fresh vision of foamy teeth coming over the rim of the barrel. After some part of an eternity I took the calculated risk of screaming for rescue. Being inside a barrel inside a room inside a barn, it was a long shot. But at least it didn't bring the fox, whose stealthy movements I could hear when I was motionless. (OK, that may have been the sound of dust motes settling.) Over the next half-eternity I screamed a few times, to no avail. When eternity reached an end, I took one deep breath and then cleared the barrel, the feed room, the barn, the adjoining shed, and the house entryway (locking the door) in a single motion.

It's probably fortunate that my older sister was fearless. She loved a good game of nighttime capture-the-flag. And in our room at night she might slide soundlessly across the floor to pounce on my bed with a howl. Out of competitiveness, not joy, I was driven to participate, retaliate, and thereby learn to confront my various terrors. Within my family, my claim to fame was wanting to go to bed on time. But thanks to my environment, by the time I reached adolescence I enjoyed an unforeseen reputation among my friends as the adventurous one. That's the impact that "nurture," or environment, can have on a personality.

Then in college a spider phobia erupted. And in my thirties a cluster of stressful events brought on the curious symptom of being unable to catch my breath. The doctor was nice about it but did gently suggest I might have a screw loose. Life brought more stressful events, as it will, and there came a point, shortly before I crawled in the door of a psychiatrist's office, where I slept only a few hours a night, waking to ruminate on the injustices of this harsh and bloodstained world, my heart speeding up, my chest heaving as I fought, fought, fought to sigh.

The drug that the shrink gave me demonstrated just how biological my personality was. Over a couple of weeks I became a different person. You'd still have recognized me. I just wasn't impatient, brooding, frustrated, worried, and heaving great sighs. I was at ease.

Meanwhile, halfway down the Atlantic Seaboard . . . a different cousin, Eleanor, was growing up to be one of those people so non-Neurotic that you almost wonder what's wrong with them. I rarely saw her when I was a kid. But I do remember the time she and her sister took my sister and me camping. It was different from when our dad took us camping. With Dad, we spent two days assembling raincoats, good socks, whistles, compasses, matches in waterproof containers, pocketknives, and other instruments of survival. When Eleanor tossed us in her car she also tossed in a tent, a box of cookies, a jar of peanut butter, and a bikini. We had a blast, climbing a mountain in our flip-flops, and leaping off a footbridge (well, the fearless among us, anyway) into a moss-rimmed pool.

When I reconnected with Eleanor a couple of decades later, the only thing that had changed was that she had conquered the world of underwater sound propagation, written computer programs to interpret it, and was so busy teaching it to submariners around the world that she's now kind of hard to catch up with. But one spring weekend my spouse and I joined her spouse and her on their sailboat in Chesapeake Bay. At this close range, I thought I might make an investigation of that unflusterable personality.

But first we had to cast things off and heave things up and grind things in and out in a perfectly penetrating cold fog. Maybe it was a cold mist. Or—yes, that's it—it was a hissing drizzle. Everyone did their tasks cheerfully except me. I have Reynaud's syndrome, so my fingers turn numb and white in cool, damp weather. I hunched in three sweaters and tried to look happy. Later when the water got lurchy, I really did help, taking almost complete responsibility for the binoculars, which might otherwise have hopped over the little wooden thingy and into the storm-tossed sea. I also commanded the rescue of a paralyzed duck who turned out to be a plastic decoy.

That night as we sat around the galley table, the waves and the rain competing to make the most watery noises, I felt sufficient blood return to my fingers that I could grip a pen. Eleanor and my similarly unanxious mate, John, lounged on the benches opposite Phil and me, who both like the shielding effect of a table before us. ("Table" seems such a nonyachty word that I must look it up in the yachting dictionary. But no, despite the fact that on a boat the table has a special rim to keep your "mess" from sliding onto the "sole," and it can often tip, or fold, or descend to become a "berth," it has managed to remain a "table." Remarkable.) Even though it was not his personality we were about to dissect, Phil began to roll a block of wood across the table, his long fingers rotating it carefully: *blonk, blonk, blonk.*

"Phil's anxiety bothers me," Eleanor begins cheerfully. "He's intelligent. He should be able to talk himself out of it. He should be rational. I can understand, and tolerate, anxiety in anybody but Phil. Because I can't walk away from him." *Blonk, blonk.*

"She," Phil informs the block, "is relaxed even when she's left her glasses behind at a hotel."

"I can buy new ones, they can ship them." She shrugs. I try to gain control of the interview.

"If I'm walking down the street and a mentally ill homeless person yells at me, I'm going to feel terrible for the rest of the day," I propose. "How do hostile confrontations affect you?"

"Really?" She is genuinely concerned about me. "Why wouldn't you say, 'Oh my god, what an ass, no wonder he's homeless'?"

She thinks about my question. "I've never worried about what people thought about me," she concludes. "Well, I don't want people to think I'm stupid. But in high school, you know when they'd write something under your yearbook picture? Question marks. That's what they wrote about me."

Blonk. "She has no need for affection."

Eleanor laughs. "You have an extra need for affection."

I break in again. "So no social anxieties. Phobias?"

"Oh! I was afraid of heights for a few years! It happened on a railroad bridge, and I suddenly had to go down on hands and knees and crawl. And then after a few years it just went away."

Blonk, blonk, blonk. "It seems like no one ever told her to be careful," Phil says. "Didn't her mom ever say, 'Don't swing that high, don't pat that skunk'?" She smiles. She ponders, trying to be helpful.

"Sometimes I can't sleep," she offers. "I can't sleep if we're fighting. And when the stock market crashed I couldn't get to sleep until we decided what to do. I also have a grooming thing—if there's a little dead skin on my thumb, I feel like I need to fix it, to clean it up." Her thumb looks a little bit gnawed on.

Blonk, blonk, blonk, blonk. Phil rolls his block away, then back again.

Eleanor grins. "I've always wanted to go to a shrink and say, 'Find something wrong with me. I dare you.' "

How can Eleanor and I be so different? We're cousins. We share the family height and coloration, and the family reflex of analytical thought. Why aren't we both anxious, or both fearless?

Perhaps the family isn't the best place to go hunting for people with similar personalities. The numbers are against you. Let's say the human genome has a total of ten genes. Number one makes hair color, number two makes eye color, number three makes anxiety, and so on. When two humans mate, each throws his or her whole genome into the melting pot. Inside the egg a new genome known as "Me" builds itself, selecting Mom's version of number one and two genes, then Dad's version of three, Mom's four, Dad's seven, eight, nine, and ten genes. A couple years later my parents toss together their twenty genes again. "My Brother" now undertakes the same random process. For each of his ten genes there's only a 50 percent chance he'll select the same version I used.

It's likely that thousands of genes, not ten, contribute to human personality. You can see how my brother and I can be 50 percent genetically identical, but worlds apart in temperament. And as cousins, Eleanor and I share only a 25 percent overlap in genes.

This math doesn't bode well for my romantic side—or rather, my fantasy of having a romantic side. My dad used to tell us that a pirate haunted the Holmes genealogy some generations back, a buccaneer by the name of Murchy. Wouldn't it be grand to be part pirate? And it would explain my interest in gold and parrots. But such is the power of dilution that distant ancestors contributed almost nothing to the DNA my parents eventually tossed in the pot for me. Even if a pirate does lurk ten generations ago, he's just one of a thousand ancestors who flipped a piece of eight into my personal gene pool. Argh, indeed. That would explain my disinterest in scuppers and yardarms.

And if the numbers don't favor familial personality patterns, then what (besides random chance) makes me anxious, and Eleanor not?

Scientists cannot drop humans into mazes to analyze our anxiety. But they have other ways to determine which of us are anxious and which are not.

A surprising number of college students volunteer for the experiments that scientists think up. This pool of guinea pigs is augmented by volunteers who have psychiatric disorders—phobias, serious depression, panic attacks, and so on.

The simplest way to identify anxious humans is with a questionnaire. It's cheap, low-tech, and easy. You just collect a hundred college students and give them the anxiety questions: Do you worry? Do you feel restless, edgy? Do you experience muscle tension?

Now you separate your top ten worriers from your bottom ten, and ask both groups more questions to tease out more differences between them:

> Are you afraid of spiders, heights, snakes, the dark, or clowns? *Voila!* Anxious people are more likely to have a phobia!

> Have you ever attempted suicide? *Sacre bleu!* Anxious people face a greater risk of killing themselves!

> Have you ever had a heart attack? *Mon dieu!* Anxious people are heart attack prone!

(All true, research suggests.)

Scientists can also combine a questionnaire with genetic analysis. That's how Lesch identified the gene he later disabled to make his anxious mice. He used questionnaires to identify anxious people; and he used genetics to determine if the anxious people were more likely to have the short version of the street-sweeper gene. *Notre dame!* The two groups indeed overlap! If the short version makes a person anxious, maybe knocking out the gene completely would make a mouse super-anxious . . .

A questionnaire plus a flock of identical twins can reveal how much your parents are to blame for your anxiety. Fraternal twins have no greater overlap in their DNA than any other two siblings. But identical twins are the product of one egg and one sperm, they share one twist of DNA. If personality were 100 percent heritable, then you would expect identical twins to have 100 percent identical personalities. They don't. If personality were zero percent heritable, then you would expect identical twins to have personalities as different as any two siblings. They don't. Identical twins fall halfway between the two extremes.

Twin studies tell us that about half your personality is your parents' fault. The results for nearly every personality factor—Extraversion, Conscientiousness, Agreeableness—cluster in the 40 to 60 percent range. Twin studies cluster so reliably in this territory that it's fair to generalize thus: About half your personality is genetically determined, knitted in DNA nine months before you were born. Half of the difference between my anxiety level and that of any other human is due to genes. The rest is explained by my environment—the influence of my childhood friends, parents, accidents that scared me, triumphs that emboldened me, and so on.

Questionnaires have taught us a lot. However, as you probably noticed when you considered the questions above, there is wiggle room in a questionnaire. Humans are poor at assessing our own personalities. Pitfalls are many. We want to appear nicer than we are, or less Neurotic than we are. We might carry an outdated image of ourselves. We're tired or in a bad mood when we take the test. So although the "self-report questionnaire" is cheap and easy, its data isn't of the highest quality. Somewhat better are questionnaires administered by a trained psychologist, who studies your demeanor and mental state as you respond to her questions.

Still, there's ample room for improvement. Science likes to measure effects in millimeters and minutes, not in "rarely," "sometimes," and "often." Measuring personality has sometimes seemed easier to do in mice than in humans.

That's changing. Down the hall from Lesch's office is one of the people who found a way to get into the human brain without cracking it open. Andreas Fallgatter is brisk, almost brusque, with hair like a brown brush, sticking straight up. Like the other PhDs at the *Nervenklinik* he wears a white lab coat. His plant collection is augmented with crayon artworks and an espresso machine. Fallgatter began his career at the same time that psychiatry was gathering itself to leap past the questionnaire.

"I could see that the work was moving in two directions, genes and imaging," he says. Lesch chose genes. "I went with imaging."

"Imaging" is the new term for what used to be the X-ray department at the hospital. The humble X-ray has served humanity long and well, but like a questionnaire it suffers from crudeness. It misses many shades of meaning. So hospitals also have CAT and MRI machines to make different images of a body's insides.

Psychiatry, too, has added tools to its kit. EEG measures electrical changes in the brain when a person talks, or solves math problems, or views a photograph of a spider. When Lesch discovered the genetic connection to anxiety, Fallgatter wondered if electricity flowed differently in a brain that carried the short version of Lesch's street-sweeper gene. And it did.

MRI tracks blood flow rather than electricity, to illuminate the brain's behavior as it solves problems and confronts spiders. Each task sends fresh blood to a different section of the brain as the cells there go to work. So that was the next test: Do people with the short version of the serotonin street-sweeper gene have different blood-flow patterns? Yes, they do. What's more, where questionnaires had winkled out only a modest connection, the MRI result was big, bold, and unmistakable. The brain, long walled off from the prodding of scientists, had been breached.

And the emerging image of an anxious brain? Its amygdala is an alarmist.

The amygdala lies roughly between your temples. (Although it's referred to in the singular, you have an amygdala in each hemisphere of your brain. It's a scientific convention to refer to the matched parts in the singular, the way you might say "the foot is connected to the ankle.") The humble almond functions like a command center for the brain in times of fear and defense.

When I open a cupboard and encounter a spider, each of a handful of my brain regions—the senses, memory, decision-making—grabs its red telephone and makes an emergency call to the amygdala: *Eight legs! Lots of eyes! Moving our way! Reminds me of a spider!*

The amygdala formulates an emergency response: *SCREAM AND RUN!*

Sometimes my logical frontal lobes can analyze the situation in time to put in a second call to the amygdala: *Never mind. I think I can just back up.*

When you show an anxious person a photo of a screaming face, her amygdala flares with alarm. When you flash a spider photo, her amygdala writhes with panic. During her days of premenstrual moodiness, a woman whose amygdala is normally tranquil flashes like lightning.

Not that you have to go out of your way to scare anxious people. From the busy world around us, an anxious person's amygdala works subconsciously to seek out the signs of danger. In a sea of human faces, an anxious person's amygdala will notice the one face that's scowling or crying. *Something's wrong! This could bode ill for us!* When reading, the anxious brain will dwell upon the words related to pain and suffering. *Take heed!* When a child pops a balloon, even if the anxious person is forewarned, the amygdala will propel the person off the ground: *Run! Get out! Save yourself!* The anxious amygdala works tirelessly to spot and react to danger.

This revelation does not come from mice. This comes from people—people with extreme personalities. And those are a dime a dozen. For someone with an anxious personality, it's just a hop, skip, and a jump to a formal anxiety disorder. I've had most of them.

Phobias are probably the best known. I don't want to sound greedy, but I have two—spiders and humans. Before I discovered serotonin drugs, a social phobia made me loath to eat in public, purchase a bra, or call a stranger on the phone.

Post-Traumatic Stress Disorder (PTSD) epitomizes an anxious amygdala's refusal to unlearn a scary lesson. Years ago I awoke to the sound of an intruder in my house. The event made such an impression on my amygdala that for months afterward I would wake at the hour of the original trauma to lie paralyzed with fear. Night after night my amygdala was wrong. But it couldn't unlearn the lesson.

In **Obsessive-Compulsive Disorder (OCD)** the amygdala seems to act like one of those birthday candles that lights itself each time you blow it out. My friend "Christina" had a typical version. When she tried to leave home for an errand, the fear of her stove causing a fire would drag her back over and over. To her surprise, serotonin drugs doused the candle for good. The same drugs worked for the "OCD mice," a gene-altered strain that groomed so relentlessly that the mice kicked the fur right off their noses. Oh, shoot. Now I realize I'm chewing away at a bit of dead skin inside my cheek. Researching this behavior, I find that cheek biting is common, and classified variously as "self-injurious behavior," "impulse control disorder," and "pathologic grooming behavior." Curses on my amygdala!

Generalized Anxiety Disorder arises when the brain exports its discomfort to the rest of the body, in the form of aches, shakes, fatigue, and so on. "It sneaks up on me," says a friend—let's call her Celexa. "I notice that I'm sighing, and taking such deep breaths that by the end of the day my ribs ache. I feel like I'm almost there, if I could inhale just a tiny bit more air, if I could just open my lungs and stretch my rib cage one more micron, I'd catch my breath. It's almost unbearable. But, of course, I just have to bear it and pray it goes away." I get bouts of the same business if I skip a dose of serotonin drug.

Panic disorder is like a concentrated form of generalized anxiety. It comes on suddenly, with symptoms such as sweating, dizziness, a pounding heart, shortness of breath, and hyperventilation. It usually comes and goes in about ten minutes. In one study, researchers found that serotonin's role as traffic facilitator was impaired specifically in the amygdala of people with panic.

These disorders are all associated with one hyperactive organ. In each case, studies that spy on a working brain find an amygdala that overreacts to life's little trials. In Generalized Anxiety, the amygdala appears to have an "anticipatory" glitch: Just telling my friend Celexa that I plan to pop a balloon in ten seconds sends her amygdala into a tizzy. The phobic person's amygdala overreacts strenuously to anything scary or disgusting, not just spiders or clowns. For the social phobic, seeing or hearing any indication of disappointment or criticism can inspire subconscious alarm. And in PTSD, a specialty of warfare, one study of soldiers found that those with an injured amygdala were somewhat immune to the anxiety disorder. No amygdala? No fear.

Anxious brains show a second abnormal region, too. Scientists generally believe that the prefrontal cortex, the part that fills your forehead, is the CEO of your brain. Like the ancient amygdala, the more recently evolved PFC receives data from many sources. But unlike the amygdala, the PFC weighs the information and forms a rational opinion about it: *The amygdala says the spider is going to kill us. But memory is arguing that this never happens. We'll let memory lead on this project. Amygdala, stand down.*

Researchers are producing two lines of evidence about the PFC's role in anxiety. One is that the PFC sometimes goes suspiciously quiet during an episode of anxiety. It's as though scary situations cause the CEO to walk into a closet and shut the door. The other line of evidence reveals that the PFC and the amygdala are suffering from a faulty connection. In this case, it's as though the amygdala can send email to the PFC to warn that a spider is attacking, but when the PFC sends a calming reply to the amygdala, the message bounces back.

In both scenarios—the absent executive or the broken email—the amygdala continues to do its job. But without the steadying hand of the executive, it doesn't know when to stop.

Of course, serotonin also plays a part in anxiety disorders. The mice told us that. But human experiments and experience add weight to the case. My own experience sums up the situation nicely: I sought serotonin drugs for what I thought was depression. A few weeks later, I found myself purchasing feminine hygiene products without so much as a blush or a stammer. The difference in my anxiety level stunned me. The psychiatrist laughed and said, "Oh, you have social anxiety. These drugs often treat that."

And the drugs treat other anxiety disorders, too, in a hit-or-miss way. My social anxiety is gone, but my spider phobia is alive and well. I'm still chewing my cheek. And when my hormones bottom out a few days each month, all the drugs in the world can't pacify my amygdala. For any given anxiety disorder, serotonin drugs help about half of patients.

I wonder if serotonin isn't especially important in social fear. I started to ponder that after I read about the social lives of locusts. Locusts are loners until their population grows to a certain density. Then, each time two locusts see each other or accidentally brush legs in passing, their serotonin levels rise. Over a few generations, they metamorphose into the most sociable of herd animals. This is when they congregate in swarms that travel hundreds of miles and do millions in crop damage. Without the serotonin, they avoid one another. A locust with his serotonin blocked by a scientist gets no kick from social contact, and goes his own way. It's as though serotonin allows locusts to let down their guard in order to function as a group. A related experiment recently allowed aggressive humans to determine the number of electric shocks their opponent would get in a computer game. After taking serotonin drugs the aggressors shocked their opponents much less. So perhaps a healthy serotonin level helps an animal—human or locust—to get along with strangers.

The reason for serotonin's fitful effectiveness may lie in the snarled connections of the human brain. Our brain is a terribly complicated organ, with newer parts hooked on to older parts with the equivalent of coat hangers and duct tape. The reality is that many brain regions besides the amygdala and the PFC get riled up during anxious moments. Each of these regions sends email to many other regions: *Calm down. Freak out! Help me. Remember this! Do we look stupid?* The traffic gets fouled pretty quickly, and scientists struggle to find meaningful patterns in the mess of messages.

The brain chemicals are similarly entangled. Serotonin is a big player, but there are many others, all interacting. As one rises, another falls and a third does cartwheels.

Before now I had not taken an inventory of my personality, let alone my anxiety disorders. I'm surprised to learn how anxious I am. I had no idea my amygdala was such a lily-liver, such a chicken heart! Nor did I realize that my assorted derangements all cower under the big tent of anxiety.

And I didn't realize how common the anxiety disorders are. Nearly one-third of all Americans will be overwhelmed by a phobia, panic attack, or another disorder in their lifetime. Women outnumber men by a wide margin. I may be a nutcase, but I blend right in.

Evolution of Anxiety

A little anxiety is a good thing, in the right context. With all the world's Extraverts and thrill-seekers climbing trees and playing with snakes, someone has to keep the risks in mind. All the time. Every minute.

The anxiety disorders, and normal anxiety itself, share a common theme: avoidance. Anxiety is a biological method of steering an animal away from a situation that might be dangerous. And if some of us avoid a little more strenuously than others, we have only evolution to blame. If the best arrangement for our species was for us to share an identical level of avoidance, as we share an identical number of legs, that's how we'd be. Human personality would be less varied.

That's not what evolution has produced. A reliable subset of the population—of both mice and men—are worriers. I would like to know why. And why are some wild mice naturally anxious, even before they encounter scientists with their elevated plus mazes? There are also Eleanor mice in the wild, who would skip down the open arms of a maze without a care. Why would evolution produce animals that appear to be overly fearful, and overly bold?

Mice can explain it best. Because mice have simple lives, it's easier to demonstrate how a range of personalities can help their species prosper. Mice in the wild need to eat, mate, and raise offspring. How does an anxious personality help them to accomplish these tasks?

Allow me to introduce my desk mice. I had asked the Petco girl for two female mice. She lifted their red igloo and from the mass of silky white bodies snatched a pair of tails: my mice. They might as well be named Black and White, or Eleanor and Hannah. Two mice have never been so different.

At home Maxi surveyed her new plastic condo, nibbled a kibble, and vanished into the shredded bank statements I had provided. She didn't come out all day. I presumed she was dying of stress. But in fact Maxi is just a very mellow mouse. Mitzi paid no heed when Maxi retreated. She had already explored the condo four times over, at four times what I considered a normal mouse velocity. She then hopped onto the wheel and began to run. She ran. And she ran. And she ran. I left the house for an hour, and returned to find her running. She had dragged a food pellet near so that she could refuel without climbing down. While running, she would lean out of the wheel as if checking her progress: *Am I back at Petco yet? No? I'll keep running. OK, how about now?*

Mitzi needed days to acclimatize to the condo. Each time I entered my office or put my hand in the cage, she jumped onto the wheel and tried to run to the sky. But as time went by she ran for shorter periods before exhausting her fear. By day four when I snapped open the condo skylight to drop in a cashew, she climbed from her bed, took a few halfhearted strides on the wheel, and then rejoined Maxi in the igloo they had built from paper shreds. Maxi, by contrast, was sitting in my palm by day four. When I entered the office in the morning her head popped up from the shreds. If I opened the skylight she scurried forward, already accustomed to me greeting her with an almond, a banana chip, a bit of apricot.

So, let's transport anxious Mitzi and nonanxious Maxi into the wild, out in my backyard. We'll let them have pups. And we'll run this experiment twice, once in a wet year, and once in a drought year. Let's see what personality is good for.

In the wet year, plants thrive and food is plentiful. Although the pups nurse heavily and drain calories from their mothers, Mitzi and Maxi needn't travel far to find food. This suits anxious Mitzi. Her serotonin and amygdala are set to discourage risk-taking. And this year, there are hardly any risks to take. Buds are bursting, insects are juicy, and seeds are rattling down like hailstones. A mouse hardly has to leave her nest, and Mitzi hardly does. After all, the farther she travels, the greater her risk of encountering a cat, a fox, an owl. That's a very simple rule for most animals: The more time you spend out and about, the higher your odds of being caught. Twitchy Mitzi stays close to home, and thrives.

For Maxi, it's not such a great year. Her personality encourages her to indulge her curiosity, even when there's no need. Sure there are seeds right on the doorstep. But what if there's something even better than seeds just past the pine tree? It's a quiet summer's night, so Maxi will amble over and have a look. Around the neighborhood all the low-anxiety mice are feeling the same way: just a few more steps, just a little farther. (Meanwhile all the Mitzis are stretching one tentative paw out of the nest, hesitating before easing the next paw forward, their little hearts pounding...) So when the cats and foxes and owls descend tonight, they'll carry away our daring little Maxi. In the nest her babies will start to chill, and start to wail, and eventually die. Cautious Mitzi's babies will grow fat and strong and carry forward the anxiety genes. Round One goes to Mitzi.

Round Two finds Mitzi and Maxi's yard much changed. Plants are stunted, insects are few and scrawny. Hungry babies are nursing as though a river runs past the nest, but the two mothers find little moisture in their nighttime foraging. Every evening a thirsty Mitzi forces herself out to gather what she can find. But there comes a night when she dares not go farther. She has sucked dry every green plant and opened every tough seed that lay within her small comfort zone. There will be less milk for the pups. They will mewl with hunger and perhaps attract a raccoon. Tomorrow the situation will be worse.

Maxi now is in her element. This lean environment rewards explorers. In her nightly adventures Maxi wanders clear around the house to where the garden hose drips—an oasis in the desert! Replete with wildlife—succulent earthworms and pill bugs abound! Maxi's babies will stay chubby and pink.

That's the story: In a rich environment, anxious mice thrive. In an impoverished environment, the mouse who hesitates is lost. From this very simple formula comes an entire dimension of personality. Every animal on earth fits somewhere on the spectrum. Some of us are quick to approach, others are eager to avoid. Some of us want to ski across Antarctica, others want to take a bus tour of Vermont's covered bridges. It takes all kinds.

Or . . . is this a "just-so story," patched together to explain a set of circumstances?

Science has a way of shaking "just-so stories" until they collapse. But the process can also uncover facts that reinforce a good hypothesis. What I'd like to see is real, solid research, demonstrating how anxiety can be a good thing, for mice or humans. Science excels at dismissing our assumptions and sensible guesses. But how on earth can science demonstrate the benefit of anxiety? Especially in humans?

You can imagine how pleased I was to find this study result: Anxious kids die less often from stupid accidents.

You know exactly who these anxious kids are. They're *not* the teens who will climb onto a bicycle with no brakes, point it downhill, and say, "Dare me?" Later, they're not the young adults who will climb into a car with a drunken friend at the wheel. Yes, these are the kids who don't play with loaded guns, don't attempt flips off the diving board, and do tell Mom when a stranger offers them candy. These anxious kids, as a statistical, scientific fact, are more likely to live long enough to reproduce. You can't ask for a much clearer signal than that. Anxiety is good . . .

. . . and bad. Those same anxious children pay the piper in middle age as chronic stress comes to fruition in the form of fatal heart attacks. Fortunately for anxious people, evolution doesn't care about late-life diseases. Because cardiovascular disease generally strikes after we've reproduced, evolution doesn't kick us out of the gene pool.

Anxiety may also confer an intelligence benefit. According to one study, a Neurotic personality has a superior "planning memory." That's the type of smarts you need to make sure you get all three children in the car, along with clothes, toothbrushes, asthma inhalers, holiday gifts, sleeping bags, and the dog, and get on the highway before rush hour. And to take the back road at Exit 77 to avoid construction. Neuroticism may also confer a steadiness of intellect, wherein the brain can make consistent, correct decisions even under distracting conditions. Perhaps the Neurotic personality is society's stern, crabby grandfather, guiding our steps to the safest and sanest, if not the most gleeful, path.

This anxiety facet of the Neurotic factor is among the biggest—most common, most troublesome to human happiness—of the thirty facets of human personality. But you can blame part of that on the fact that we have altered our environment so radically in the past few thousand years. If we return to the assumption that anxiety is about avoiding danger, then we all look a little less crazy.

Perhaps in a quieter environment than this modern world, we avoiders wouldn't have our nerves jangled so often. If transported to a simple cave life, we might be admired for screeching when we spied a snake in the trail. We might be respected for our cautious approach to strangers. Our hearth might be the most popular spot to eat because we washed our hands so often that our food rarely made you vomit. Our avoiding ways, currently derided as timid or hysterical, would look as reasonable as the conservative behavior of our Mitzi mouse. We'd be the clan members who were never sidelined by the infected bite of an inappropriate pet. We'd never fall out of trees while reaching for that fruit that's just a tiny bit out of reach.

Yes, we would face starvation when the fruit trees failed. But that's why diversity is good. That's why I like to stay in touch with Cousin Eleanor.

THIS GIVES YOU A QUICK look at where you land on this facet. If your answers tend toward the "often" side, you're higher in that facet.

The depressive personality has a tendency to look on the bleak side. If you score high on depression, then you may tend toward the skeptical, solitary, and slow-moving end of this personality measure. That doesn't mean you're always blue. You just aren't going to be voted "most lively and optimistic."

If your score is mid-range, you might not always think that you, or the world, are in great shape, but neither are you hanging your head like Eeyore. If your score leans toward "rarely," then you probably have a solid respect for yourself, you feel motivated, successful, cheerful, and you're eating and sleeping like a champ.

Then again, Klaus-Peter Lesch, whose mouse lab I visited in Germany, isn't so sure there is such a thing as "real" depression. He's met plenty of people who feel hopeless, lifeless, and spiritually spent. Both he and his wife treat these patients in the psychiatric hospital downhill from his office. Sometimes they can give a patient a pill and drive the gloomy feelings away. But other times they try pills, and talk therapy, and different pills, and still the patient suffers.

"I don't believe any two patients have the same depression," he proposes. "Depression is a group of disorders. There could be a hundred or a hundred and fifty disorders within the framework of depression."

Wow. That raises an interesting problem, doesn't it? The people who study the biology of personality are discovering that we're too complicated to describe. And yet the best tool psychologists have to diagnose us is a very simple list of descriptions. The famous *DSM* (*Diagnostic and Statistical Manual*), the book that tells you whether you're officially off the rails, is not based on scientific research. It is not based on discoveries about serotonin and genes and neurons. It's not even based on studies of huge numbers of people, which might have revealed subtle patterns. First published in 1952, and based on an army manual, the *DSM* creates, out of thin air, names for the various ways a human personality can go haywire. The scientific basis of these invented categories is ... well, it's mainly based on how psychiatrists view their patients. So it's really not very scientific.

And that makes it difficult to come up with a logical treatment plan for depressed people. Talk therapy works for some, but not all people. Serotonin drugs work for about half. Some people recover spontaneously, others never rise from depression. Clearly, we don't understand this disorder very well. So we look to the mice.

Depressive Mouse

Because human depression is so poorly understood, scientists have created oodles of depressed mice. It's instructive that this is even possible. That there are so many ways to lower the energy level of a brain tells us a couple of things: One, it may be a common condition in wild animals. Two, as such, it's probably not a dire, horrible fate.

But first things first: Given our confusion over what depression is, how would you know if you succeeded in creating a legitimately depressed mouse?

Just as the elevated plus maze is a standard measure of mouse anxiety, the forced swim and the tail suspension are common tests of murine moroseness.

For the first, a researcher lowers a mouse into a beaker of water. The mouse swims around the beaker for a while. She might try to power herself up the side of the glass. She will probably poop a few times. (Mice do that for any new experience. Mitzi and Maxi used to do it when I handled them.) But after just a few minutes the average mouse will quit. She'll float quietly. (She will not drown. Mice are expert little swimmers, and can paddle for two and a half hours. They're also good floaters. When they stop swimming their faces rest above water and they appear calm.)

Regardless of how it appears, researchers categorize the floating as "despair behavior." I expect this is a misnomer, since *despere* is Old French for "lose hope." Hope implies a sense of the future. Mice aren't likely to have a sense of the future, hopeful or otherwise. They live in the moment. And if at the moment they are stuck in water, then *que sera, sera.* Given razor blades, they wouldn't slit their little wrists because they don't realize that their future is bleak. So the term "acceptance" might fit better than despair. Whatever you call it, the mice who quickly substitute floating for swimming are considered to be depressive, and therefore useful for research.

To conduct the tail suspension test, a researcher sticks tape to a mouse's tail then hangs the animal from a tiny scale. The scale records the mouse's struggle to get free over the course of six minutes. To minimize distractions, this experiment often takes place inside a small white cubicle, with data flowing out to a hungry computer. (I admit I don't like lifting Mitzi and Maxi by their tails, but it really does seem to be fine. Veterinarians do it. Even people who run websites dedicated to pet mice do it. I can find no indication, anywhere, that it hurts them. In fact, they use their tails as some monkeys do, gripping with it as they climb. But lifting them by the tail just seems rude.) As with the forced swim, it's the time a mouse spends hanging motionless that scientists measure. The ones who cease struggling soonest are the ones you want to test drugs on.

So that's how scientists determine if a mouse is depressed. The forced swim and tail suspension tests are the most common in part because they are the least distressing to mice.

If that were the only indication of mouse "depression," I would have my doubts. It seems to me that the first mice to quit swimming could also be considered the most conservative ones. Accepting that their environment is currently impossible to conquer, these mice save their energy. When their environment changes, they'll have the strength to resume their tasks.

But the whole point of creating depressed mice is to find drugs that combat depression in people. So what happens when you feed a despairing mouse a nibble of serotonin drug? She swims longer before conceding to her environment and wriggles longer when hung by her tail. This is reassuring. It means a mouse is on the right track. It can be used to try out new drugs that have fewer side effects or a greater success rate.

What, then, do the legions of despairing mice tell us about our own personalities?

Curiously, one of Lesch's depressed mice has demonstrated that too much serotonin in the synapses can bog a mouse down just as effectively as having too little. Such an overloaded mouse, Lesch found, displays classic despair behavior. When hung by the tail, she gives up the fight.

In mice, as in man, this is contrary to the original view of serotonin levels and depression. The whole point of serotonin drugs is to increase the serotonin in the intersections between your neurons, accelerating your neural traffic. That being the case, you might think that too much serotonin would make a creature feel a bit busy in the brain, if anything.

But in mice, extra serotonin doesn't appear to cause euphoria. It makes them hang by their tails, it makes them quit swimming. And if this is the case with mice, it may well be true for us. That could help explain why serotonin drugs are ineffective for so many depressed people. If there's already too much serotonin in your intersections, then slowing the rate at which your street sweepers remove it could compound the problem.

If this does prove true for humans, what would it reveal about normal human personality? That a person who leans to the depressive end of the spectrum might carry either a little less or *a little more* serotonin than average.

Interestingly, the excess serotonin in Lesch's depressed mice may have changed the way their brain cells grew when they were forming. In these mice some neurons in the amygdala, the emergency center that's so alert in anxious people, branch out to form many more synapses than usual. Lesch thinks this might result in an amygdala that, as in anxious people, is in a perpetual state of emergency. A second deformity strikes nerves in the mouse's prefrontal cortex—the brain's analytical region. Here some neurons grow branches that are much longer than normal. Lesch thinks they may be responding to the serotonin surfeit.

The implication for us? Brain wiring happens very early in the life of mouse and man. Judging from the mice, if your serotonin is misadjusted at that time, it could influence the architecture of your brain. You may have patches of odd or hyperactive neurons for the rest of your life. As a result you might be quick to give up when times are tough. If your serotonin is perfectly average when the wiring is laid down, you could end up with textbook neurons. You would be emotionally stable, positive, and resilient, even when hung by your tail.

Because this speculation is based on mouse research, we can't be sure that the brain of a depressed human would develop the same glitches. The beauty of mice is that you can, after consulting your conscience, dissect their brains and make discoveries that may lead to happier, healthier humans in the long run. Case in point, regarding Lesch's high-serotonin, depressed mice: He has found a drug that, when given to them as babies, as their brains are developing, restores order. They grow up to be mice who will swim and wriggle as optimistically as any other. A drug isn't allowed to cross from mice to humans until it undergoes many years of testing, but this is how they get started.

Depressive Human

The human head, too, has given up some secrets about the biology of the depressed/cheerful spectrum of personality. The mouse brain may offer superior accessibility, but the human brain shines in communicativeness. Example: When a dog is sick he lies silently on his bed. When a child is sick, she says, "My tummy hurts right here."

Lesch learns a great deal from his mice, but he also goes right to the human's mouth. Since his discovery of the long and short versions of the street-sweeper gene in 1996, he and dozens of other researchers have looked at how the two different alleles of that gene may affect human personality. In a nutshell, if you've got a short version, you're more inclined toward vigilance and avoidance—or as the psychologists phrase it, anxiety and depression.

For example, one study found that girls (but not boys) who carry the short version and are abused as children have a higher risk of tilting into major depression as teens. Rather than striking back at a threatening environment, those girls carrying a short version are more likely to withdraw. Furthermore, research has found that among clinically depressed people of all types the short version is more common.

Shorties are also more likely to be ruminators when they're under stress. If you don't know what this means, aren't you lucky! I'm all about ruminating, if I forget to take my serotonin drug. I can pass hour after hour rehashing a situation that makes me feel rejected or worried. On the anxiety front, Shorties have jumpy amygdalas, as measured by a "startle response." They're less able to damp down their fear. And on the whole a Shorty's amygdala appears to burn hot.

Lest you fret: This list of tribulations borne by the Shorties may look like a crushing burden. The truth behind the headlines is that these studies turn up only tiny differences between people with the long or short versions of the street-sweeper gene. Some studies find no differences at all. Using statistics, scientists can winkle out little trends that grow into big headlines. But if you lined up your friends and tried to guess who were Shorties and who were Longs, you'd fail.

One hard lesson of Lesch's famous discovery is that it takes more than one gene to make you happy or sad. Lesch estimates that this one gene explains only 3 percent of the difference between my Neuroticism and yours. That means many other genes chip in their two cents' to your Neuroticism, whether you rank high or low. Lesch says the number of genes may be in the hundreds, assuming that the impact of his street-sweeper gene is unusually large.

This is inconvenient. Early in the history of human genetics, researchers hoped to pin each human attribute—large ears, small stature, giddy Extraversion—on a single gene: one trait, one gene. No such luck. Researchers now realize there could be dozens or hundreds of genes involved in a feature like human Extraversion or Neuroticism. That being the case, many are looking for simpler ways to size up the biology of a personality.

These days, one of Lesch's closest collaborators works in an office four thousand miles away. The distance matters little. They're scientists. The Internet is an extension of their fingers. Even as I was talking with Lesch one day he excused himself to dial up his colleague on Skype, the program that downloads friends and colleagues right into your computer, voice, face, and all. On the other side of the Atlantic Turhan Canli was not answering.

Back in the United States, I, too, had some difficulty pinning him down. He is, Lesch had warned me, one of the busiest, brightest people he knows. When I finally got a date to visit his office at Stony Brook University on Long Island, I got a last-minute email reading, "Are you allergic to cats?"

"Only the ones that eat songbirds," I jested.

"Then we're good."

The walls of Canli's office are cherry red and yellow, a shocking departure from the dirty beige wastelands of traditional academic decor. The bookshelves are of curving metal. Also highly irregular. Big abstract paintings hang on one wall. Odd! And the cat, a sleek silver beauty with chiseled cheeks, watched me from under the desk. The scientist was similar, but in a chair.

Canli is Turkish and German, but hard to place. The olive skin and coarse hair are consistent with the Turkish style, but high cheekbones and a broad nose run contrary to the type. His accent is vaguely British but shaped by too many influences to trace. The cat, recovering from my intrusion, steps into Canli's arms and settles with his head against the black turtleneck. He has feline leukemia, Canli explains, and can't consort with the cats he and his wife have at home. All signs of the cat's disease have vanished since he took possession of the office.

Canli looks too young to have done as much as he has. In 2001 he published the first scientific paper to link two different personality types—Extraverts and Neurotics—with two different patterns of blood flow inside the brain. He has also investigated the learning process in rats; the effects of estrogen on the female brain; how genes and environment interact to produce rumination; and the molecular basis of personality, among other things.

His reputation is growing. Quiet little Stony Brook, squatting in a patch of oak forest somehow spared by suburbs and highways, now has an MRI machine of its own. Previously, Canli and anyone else who wanted to spy on a working brain had to go to Yale, a few hours away. But Canli's work has been so productive that the National Science Foundation decided he should have an MRI machine in his backyard. It sits in a new cube of masonry a stone's throw from his office. Soon a line of researchers will form as others congregate with their own experiments.

The MRI revolution is great for researchers, if a bit weird for subjects. For starters, the tube is not much bigger than a human lying on her back. As you lie on a stretcher you can watch your feet glide into the white cave. When you come to rest, the snowy ceiling is a few inches from your nose. By rolling your eyes you can still see your feet, and perhaps a fan that's blowing air in. (The whirling magnets heat up the tube.) Now the researcher can subject you to various torments and watch your brain respond.

How do they deliver torments to you, constrained as you are? Most psychology experiments today are designed so that you sit at a computer terminal, either watching images or responding to cues and challenges. Scientists capture your brain's electrical activity with an EEG cap, or give you a button box with which to click your responses.

Not even the slimmest of computer monitors would fit between your nose and the ceiling of the MRI tube. But scientists have adapted their computer tests to the tubular environment: Using mirrors, they bounce the image of a computer screen into the tube. Now they can beam you the image of a spider, or a screaming baby, or whatever they're using to stimulate your brain. They snake a button box in to your palm. It's as though you were sitting at a computer, taking the classic tests, but you lie verrrry still and scientists spy on your darkest secrets.

Canli thinks that the MRI machine knows us better than we know ourselves. Before MRI, scientists had to take our word for what kind of people we are based on questionnaires. Now they're discovering that the real story is under the skull. The blood and guts of the brain cannot tell a lie. Blood rushes to those regions that react to a given stimulus. You can't help it. And the MRI machine can see it. So let's go back to the street-sweeper gene. Let's see what an MRI says about people with the short version versus the long version.

Here's a classic experiment: A group at the National Institutes of Health (NIH) assembled a sample of humans and checked their street-sweeper gene for short and long. Although they were studying anxiety, not depression, the process is the same for any feature you might want to investigate. The subjects answered a questionnaire about their own anxiety. Merging this data revealed no link between the short version and a fearful personality. Then scientists ran the people through an MRI test. Under the guise of asking them to match up facial expressions, they beamed photos into the tube. The subjects proceeded with their fake task, matching happy face with happy face and scary face with scary face. The researchers eavesdropped on the amygdala.

Shorties may not see themselves as anxious, but their amygdalas said otherwise. Each time a scary face appeared, the Shorty amygdala reacted with five times the force of the Long amygdala. That's a whopping difference. In a traditional experiment, based only on a questionnaire, the short version usually appears to make only a few percentage points of difference in how anxious a person is. That's on a good day. Often such experiments find no link at all. On a bad day they find a negative link: The short version appears to make you *less* anxious. Differences this tiny cause massive headaches for scientists. MRI is a massive aspirin. It bores through the skin of a person's self-image, and goes straight to the heart—brain—of the matter.

Scientists are also discovering structural difference in the brains of Shorties—differences in where gray matter concentrates, or in how densely the neurons are packed. Canli, by looking at the entire brain at once, found that the short and long versions predict how much gray matter will reside in various regions. He can't yet tell what that portends, for depression or anything else, but it's a difference that invites more research.

Often these MRI experiments look at the large factor, instead of the smaller facets of personality. Lumping depression and anxiety together shifts the focus to that old approach-avoid spectrum.

Some depression researchers are zooming in on the venerable amygdala. They've learned that the left amygdala (remember we have two) tends to be the more active of the pair in non-Neurotic people; the right is more active in those of us who rank higher on anxiety, depression, and other Neuroticism facets.

Boring deeper, Canli and his colleagues have used MRI to measure the density of those amygdala. Again what they found was enigmatic: Extraverts, who are socially eager and outgoing, have more gray matter in their left amygdala. And Neurotics have less gray matter in their right amygdala. His friend Klaus-Peter Lesch found a similar connection between the amygdala and people carrying the short version of the street-sweeper gene. Canli says this fits a proposal that people with high Neuroticism might be "right-brained" in general; and that non-Neurotic people are more "left-brained."

For the moment no one is sure how all these physical peculiarities im-

pact your personality. But the structural differences add to the understanding of how personality is built, chemical by chemical, cell by cell.
As the research piles up, the amygdala starts to look like the mascot of the
Neuroticism factor, the spider at the center of a web of brain regions that
urge us to approach or avoid.

In fact, Canli has proposed a fresh theory about the old amygdala. Perhaps a Neurotic person's amygdala doesn't overreact to signs of trouble,
he offers. Perhaps it's hyperactive right from the start. He observes that in
the MRI tube a Shorty amygdala is more active than a Long amygdala
even when it confronts a neutral photo of a spoon or a pair of shoes.
Maybe a Shorty amygdala is always poised on the edge of its chair, always,
always on the lookout for danger. When you add the normal response to
trouble on top of that, the amygdala looks panicky indeed. Perhaps, Canli
suggests, we who tend toward anxiety and depression have amygdalas that
search the environment tirelessly for signs of trouble. In other words, if
you're feeling blue or anxious, you're not high in Neuroticism. You just
have a low-density, right-handed, hypervigilant amygdala.

As a result, you may be more conservative with your joy and more
generous with your doubt. Not to a degree that's offensive or difficult.
You're just not as bubbly as the next person.

However, scoring high on the depression facet does put you at risk for
a few other conditions. One is anxiety. The depression and anxiety facets
often go hand in hand. Scoring high on one predicts that you'll score high
on the other.

Your other risk is for true depression disorder. "The Big D," as a friend
calls it, is a big deal. Major depression (MD, as the pros call it) makes millions of humans feel dismal day after day after day. The sadness, lack of
energy, and low interest in others are rough on humans, who are naturally social animals.

And once you do slip into MD, you may have trouble getting out. Just
as an anxious brain is biased to find signs of danger in the environment,
so does the depressed brain hunt for signs that times are hard, hard, hard.
You may have noticed this if you've known a person with MD. When he
reads the newspaper he dwells on the stories of abused kittens, bloody
war crimes, heartrending job layoffs, cynical politics. It looks like wallow-
ing. And perhaps it is. But it's not intentional. It's that damned amygdala.
The average person's amygdala flares to life for about ten seconds when
it's shown something sad or bad, one study found. But a depressed per-

son's amygdala flares and then continues to burn bright long after every-one else's has relaxed.

Depression, in turn, increases your risk for other diseases. Major de-pression is linked to diabetes, heart disease, osteoporosis, and stroke.

Burdensome as MD is, we hardly have a handle on it. We hit it with serotonin drugs, but a substantial percentage of people get no relief. Newer drugs that target both serotonin and a second neurotransmitter, norepinephrine, still leave many people unrelieved. These people are forced to try drugs with more side effects, and even electroconvulsive (shock) therapy. And if none of this helps, their options are to soldier on in a cheerless world or end their lives.

Anybody, from any perch along the spectrum of depressiveness, can take a tumble into the abyss of major depression. But research shows that people whose personalities stand on the more depressed end lose their footing more easily and fall deeper into the pit of despair. I do it every autumn.

Checking my autumnal funk against the *DSM,* I am disappointed to learn that Seasonal Affective Disorder, like my collection of anxiety dis-orders, also qualifies as a formal malfunction. Here is the *DSM* list of nine potential Depression symptoms:

- changes in appetite

- changes in weight

- changes in agitation

- changes in sleep patterns

- morbid thoughts

- feelings of fatigue

- feelings of disinterest

- feelings of gloominess

- feelings of worthlessness

To pass the test, you must experience at least five of the nine, daily, for at least two weeks. Those five must include either gloominess or disin-terest.

That's vanilla major depression. But MD comes in different flavors, to suit each personality. The most popular versions are:

Seasonal Affective Disorder   When the calendar says it's time to gear up for the December holidays, my brain says it's time to nap. I cannot find the will to leave the house after working all day. It's dark out there. I'm sleepy. Can't open the box of greeting cards. Can't hold the pen. The idea of polishing the silver and vacuuming spawns a slow marathon of yawns.

SAD seems more strongly related to climate and genetics than the length of winter nights. Like major depression, it's more common among women than men. Unlike major depression, it comes to call as the days grow shorter in autumn, and then rises and flutters away with the longer days of late winter.

Premenstrual Dysphoria Disorder   PMDD is my other depression disorder. When the hormones plummet prior to menstruation, some women become fussy. I ruminate over ancient histories, I weep over spilled milk, I mistake my husband's absentmindedness for an impending divorce. That's PMDD. Peeping into the brain, researchers have found that this version of depression dulls a woman's ability to read facial expressions and turns her amygdala to a paranoid fearmonger.

Baby Blues   This is a mild form of the better-known postpartum depression. It's tremendously common, touching 40 to 80 percent of new mothers with sadness, mood swings, sleep disruption, irritability, anxiety, and impaired concentration. It often lasts only a week or so.

Postpartum Depression   This is depression on steroids—or hormones, more aptly. It brings severe mood swings, social withdrawal, overwhelming fatigue, anger, and inability to bond with the baby, along with thoughts of suicide and infanticide. The babies in question face an increased risk of personality disorders of their own, perhaps due to that lonesome start.

But again, these are forms of depression—serious disorders. The fact that you have a depressive personality is no guarantee that you'll ever experience a major depression. You just err a bit on the cautious and pessimistic side of life.

Evolution of Depression

The personality type that leans toward depression must be useful to both mouse and man. Were it not, evolution would make it go away. So what's it good for? This isn't the kind of puzzle researchers contemplate. They focus on finding ways to help people who have plunged into serious malfunction. The evolution of such a personality type is not their concern.

The puzzle is beyond my expertise, too, but I'm going to venture an argument. I'm going to suggest that a depressive person is a mouse who's quick to realize that her environment is currently stacked against her. She's a mouse who opts to hoard her energy until things turn around.

Let's set aside the cultural baggage we heap onto depression. In many cultures a depressed human is considered a disappointment. In a busy-busy context, a depressive person may look lazy. In a culture that treasures social interaction, a depressive person may look unfriendly. These judgments aren't lost on the depressed person. She's already inclined to view herself as a waste of space. When her culture sneers at her and her friends abandon attempts to lure her out, her sense of worthlessness is confirmed. She's in full agreement with the cultural diagnosis: She's a loser.

The clash between a depressive person and her culture can make depression look more complicated than it is. So let's look at the biology of this state, free of the social commentary. Let's see how simple it really can be.

Let me build a little context.

Scene 1: I inhale a cold virus, which founds a dynasty inside my nose. This microbe could multiply with such determination that my nose and throat become raw and inflamed. Bacteria could find a foothold in those ravaged tissues, starting additional infections. My sinuses, my bronchi, my very lungs could fester. I could die. But my body has other plans. As the environment—bacterial and viral in this case—pummels my mucous membranes, my body cancels all unessential activity. I feel leaden, boneless, sleepy. Digestion takes energy. I don't eat. Walking takes energy. I shuffle. Reading is effortful. Even watching TV taxes my torpid brain. Beautiful music can't move me and great art seems not so great. I curl up on the couch. I drift.

Scene 2: The woodchuck in my backyard eats steadily through the long summer days, storing her extra calories in the form of fat. As autumn kills off the violets and the clover she relies on, her diet will narrow.

Her body knows what the environment will do next—it will dump snow all over the yellowing remnants of her salad bar. That's a fight she can't win. And so she disappears from the lawn. Underground, she curls into a sphere, the universal posture of energy conservation. Every cell in her body adopts an attitude of frugality, burning less sugar, shedding less heat. Her blood cuts back on deliveries of sugar and oxygen to the cells. The heart slows. The lungs rest.

What we see in both scenes is an animal whose environment has so abused it that resistance is futile. The path to survival lies in surrender. The body automatically powers down. Only the vital systems are supported. When the environmental conditions improve, both animals will slowly return to life.

Now let's bring in our depressive person. Maybe with the woodchuck and me as context she won't seem such a lost cause.

A teenage girl adores her older brother. One day out of the blue the news crashes into her adolescent brain: He has died. She mourns his loss with the rest of the family. And when they slowly rise out of their mourning depression and resume their former level of activity and hopefulness, she doesn't. With her attention fixed inward, she's not enlivened by school sleepovers or family festivities.

Objectively, the depressed person could be mistaken for me-with-a-cold. She's physically slow, socially withdrawn, and numb to the world's delights. Many depressed people actually walk more slowly, and in shorter steps, compared to cheerful people. So perhaps depression, whether it's mild or severe, is an energy-saving state that humans enter when our environment proves intolerable. It allows the animal to conserve resources until conditions change.

The environmental circumstances that trigger depression are fairly intolerable. The death of a family member is a good indicator that your environment is unmanageable. Divorce, which epitomizes social banishment, is another common trigger. The loss of a job and money, which eases so many of our relationships with our environment, is another. And there's nothing like bodily losses—a heart attack, losing a leg in warfare—to make a human feel helpless.

For women with postpartum depression, an environment devoid of social support appears to be a major risk. Here's a mammal who has just given birth to an infant so spectacularly helpless that it needs her attention nearly around the clock. Without help it may be impossible for both mother and offspring to survive. In an impossible situation, some animals concede defeat sooner than others.

Children risk depression if their environment is too unpredictable or too hopeless. That might take the shape of parents whose behavior is corrupted by addiction or mental illness. Or a home so crowded that the child is rarely noticed. Or a school where bullies attack day after day.

Perhaps different types of depression serve different biological purposes. Seasonal Affective Disorder reminds me of the seasonal cycles I see in many animals who share my northern environment. The woodchuck takes it to the extreme as her food vanishes, plunging into a state of inaction so deep that she doesn't even dream. But the chipmunk, who modifies his environment by excavating holes to store his food, merely grows sluggish. For days on end he sleeps, rising only to stoke the fires of his metabolism with a meal of maple seeds and acorns. In animals, this energy-saving mode is called torpor.

Like the chipmunk, we with SAD do not lose our appetite. *Au contraire.* We stoke those fires, letting the heat of metabolism warm our bones. And SAD doesn't typically involve feelings of desperation or despondency that haunt some forms of depression. SAD mainly makes a human feel like sleeping until 10:00 a.m., eating, then napping, eating some more, then returning to bed at dusk.

This human torpor strikes me as a reasonable response to an unreasonable environment. In northern climes prehistoric humans had to get through winter with stored food. Food storage is a dicey proposition, for any animal. Others are always eager to withdraw what you deposited. Before the industrial era, a human in the north was at high risk for starvation. Fortunately, farming people didn't have much busywork in winter. Your animals were stabled and living off stores of dry grass. After you had chopped some firewood and knitted everyone a new pair of socks, why not dial back the activity level a few clicks?

(Today, obviously, the "why not" is that your friends and neighbors will think you're lazy, boring, unfriendly, and so on. Adopting a mild state of hibernation is not socially acceptable in the era of year-round activity. Going semi-woodchuck poses a risk to your popularity and your employment.)

By contrast, major depression seems more like my cold-virus response. It looks like an extreme response to an extreme environmental emergency. Whereas SAD might slow you down a bit, depression that involves weight loss is a more grave withdrawal for the entire system. This puts me in mind of the wolf who feeds from a poisoned carcass and curls into a ball and doesn't move for four days. It's an emergency measure that might save your life, or it might not.

Have I made a case that some human depression is an energy-saving mode? Is there an optimism in depression, whereby the animal is saving her strength for a brighter day?

One of the better arguments I can muster *against* myself is the self-perpetuating aspect of major depression. Although it can be triggered by the environment, it doesn't automatically resolve if that environment becomes benign. Like PTSD, major depression can acquire a momentum of its own. A depressed brain has a preference for negative input.

Whether my hypothesis holds water or not, it does simplify matters to get rid of the cultural view that depression is a failure of character. Depression is biological, no matter how it is brought on, or how it serves the animal.

As for the merely depressive personality, the normally depressive personality, what has that to recommend it? Well, why wouldn't a healthy personality also benefit from saving energy? It's not a bad policy for an animal, human or otherwise, to minimize her energy use, as long as that doesn't compromise her health. The depressive personality doesn't spend energy chasing flights of fancy. She doesn't joust at windmills or expect to squeeze blood from stones. When she does exert energy she does not waste it on pipe dreams and whims.

Like the anxious person, the depressive person retreats from risks, but the depressive person is more dubious than afraid. Hey, it's a dubious world. Go ahead and jump in if that suits your personality. But it doesn't suit the more depressive among us. You go ahead. We'll bide our time. We'll wait and see.

So You Think You Might Be Neurotic

A Neurotic personality is not a sentence to a cringing and despondent existence. Not at all. It does convey a cautiousness, a carefulness. But that doesn't rule out a life of good cheer and comfort. Take it from the mice.

The first day I entered the mouse lab in Germany I was struck by the hiss of hundreds of skittering toes. Although the clock said it was past mouse bedtime, each little black critter was nosing about its condo, nibbling a bit of food, rearing to sniff the air. Well, they were gene-altered to be anxious, I thought. They had concerns. They could smell a stranger, and that stirred them to gather information.

The second day I entered, I was more starkly stricken: Silence.

Seriously? That quickly they had accepted my voice and scent? They had laid their worries to rest that deftly? Yes. This time, each little mouse was snuggled in her igloo, piled cozily among her friends, more comfortable than curious. That's hopeful, isn't it?

Even Mitzi, my anxious desk mouse, has mastered her amygdala. I didn't realize when I handed over two dollars for her that she was bred to feed pet snakes. Or that her careless pedigree nearly guaranteed that she'd be a wreck, mentally and physically. But Mitzi's environment went from Insane Asylum to Club Med when she got to my office. It gives me great gratification to note that she doesn't even get out of bed for treats anymore. There are mornings when the condo is so quiet that I fear the girls have died, and I dig for them in the shredded paper. They both wake, squint at my fingers, and curl back to sleep.

This is also possible for the Neurotic human. If we can commit to giving ourselves the exercise, good diet, and stress management that reliably calm a brain, we can ease our personality a few clicks down the Neurotic spectrum.

Klaus-Peter Lesch surprised me when I gathered my gear to leave his mountain kingdom. I'm not saying Lesch is an anxious mouse, but he does carry a fair amount of tension around with him. As I stood to go, he suddenly asked, "Don't you want to know what causes resilience?" As though that's why I had come. As though I might want to do something about my irksome personality, beyond comprehending it.

"A work-life balance," he said emphatically, ticking a finger. "A regular lifestyle. And lots of exercise."

There you have it, from one of the foremost authorities on anxiety.

He and his wife, Silke Gross-Lesch, take me to dinner high in the vineyards at night. Lights twinkle on the ramparts of an old fortress across the river valley.

"That church, downhill from the fortress," he says pointing to a glimmering light. "A crazy woman had a vision of Mary or something, so they built a church there. The Bible is full of psychosis."

Well, it takes all kinds. Without the crazy woman, there's no hallucination, no pretty church.

"Have you read the book of Mormon?" Silke adds. "One long psychotic episode."

No Bible . . .

Over dinner Silke, whose necklace twists like silver DNA, listens intently as I continue to interrogate her husband. And during a lull she leans gently toward him.

"You really must talk to your son about this."

He looks up, surprised. "Well, he must ask."

"You know him," she coaxes. "Just talk to him."

"I think that I am boring people if they don't ask," he protests.

"What you don't know," she pleads, "is that you are the *least* boring when you are talking about things that you're interested in!"

He sighs. Tugs an ear, examines his plate. If he were a mouse he'd be skittering around the room and rearing to sift the air for additional signs of danger.

"You know, one of the most interesting things about being a biologist is seeing these children," he says, dodging, as it were, into a closed arm of the elevated plus maze. "There's a piece of me in this one, and a different piece of me in that one . . .

"The Riesling you get in the United States is bad," he continues. (Other closed arm of the maze.) "Do you want to try something special?"

I really don't. But I, too, am an anxious mouse, and fearful of giving offense.

Ice wine is made from grapes that are still in the pink of health when the first freeze hits them in winter. This is a rare occurrence. Inside the skin some of the water crystallizes, leaving a concentrated pool of the sugars and chemicals that make a grape interesting. You squeeze them before they thaw, and ferment that nectar.

The liquid in the glass is as clear as the night sky over the river, but gold. Truly golden gold. We raise our glasses. I'm an anxious mouse. I expect bad things.

It is certainly sweet, but it is so much more than that. The honeyed florals and fruit hang in balance with the acid, and a steely mineral finish leaves me wanting more.

There's more to a personality than chemicals. Environment matters. And these grapes, perhaps bland at birth, emerged from their trials as elegant and complex beings indeed.

2
FACTOR:
EXTRAVERSION

EXTRAVERSION FACETS
Impulsiveness/Novelty Seeking
Activeness
Cheerfulness
Assertiveness

THE CAST

Mascot . . . . .Madcap Mouse

Neurotransmitter . . . . .Dopamine

Brain Regions . . . . .Prefrontal Cortex,
Sensory Cortex, Motor Region, Nucleus Accumbens

EXTRAVERSION

EXTRAVERSION IS THE PART of your personality that propels you forward. If the Neuroticism factor describes the impulse to avoid, to retreat, to put on the brakes, Extraversion is the gas pedal. Extraverts charge forward into life. They're energized by other people. They're optimistic. They're naturally resilient. And they learn from life's rewards, not its punishments.

Impulsive people are the mascots of this factor. Some people call this factor novelty-seeking, excitement-seeking, and sensation-seeking, but they're all talking about the same thing. If you're that, then you are a go-forward person.

You go forward largely because your dopamine system is set a little bit differently from the norm. That neurotransmitter urges everyone—and everymouse—to go forward enough to capture the food and water and mates that survival demands. In an Extravert, dopamine urges more aggressively.

At the same time the prefrontal cortex (PFC) eases its foot off the brake. Normally the thoughtful manager of all the other brain regions, the PFC in an impulsive person sometimes seems to be asleep at the wheel.

Part of Extraversion's charm is that these brains don't pay much attention to negative feedback. So the snarls and growls that grumpier humans are liable to issue bounce right off these personalities. Spirits undampened, they come back for more. In contrast with Neurotic personalities, they're buoyant, optimistic, and up for a good time.

The social facets of Extraversion get to the heart of a social animal's dilemma: We must cooperate and keep the peace; but we must also watch out for our own interests. This dichotomy will get a full treatment in the Agreeableness factor.

IMPULSIVENESS/NOVELTY SEEKING

IMPULSIVENESS INDICATORS	RARELY	SOMETIMES	OFTEN
I SPEAK WITHOUT THINKING	❏	❏	❏
I BUY THINGS THEN REGRET IT LATER	❏	❏	❏
I GET BORED WITH ROUTINES	❏	❏	❏

THIS GIVES YOU A QUICK look at where you land on this facet. If your answers tend toward the "often" side, you're higher in that facet.

The novelty-seeking and impulsive personality thrives on a steady stream of stimulation. Quiet contemplation, to this personality type, has all the appeal of a colonoscopy. These are explorers, and as their attention skips across the landscape of people and ideas, they can seem fickle and absentminded. This personality notices everything and often reacts without pausing for analysis.

People low on novelty seeking and impulsivity are... boring. In a nice, reliable way. They don't play the lottery and they wouldn't take their thirty best friends on a cruise if they won. They nonetheless make ideal friends, because they make choices carefully and stick by their decisions. Of course, impulsive novelty seekers make great friends, too, because they're a lot more fun.

Impulsive Mouse

Marc Caron is not an Extravert. He is a cell biologist at Duke University's Institute for Brain Sciences, and he built the mouse version of my husband. Caron ignored my first two emails asking permission to visit. He ignored a third, six months later. One day I called his office shortly after 5 p.m., when secretaries tend to have left.

"You caught me," he sighed when I identified myself. "I'm playing hard to get." He had a powerful French-Canadian accent.

"I'm playing hard to avoid," I replied. "When can I visit?"

But this conversation, and subsequent ones, turned to my identity. How, he asked, could he be sure I wasn't an animal rights extremist bent on ruining his reputation or his family? It's a legitimate problem for researchers who work with animals. Or at least those who work with mammals. Animal rightists seem inured to the agonies of the world's most popular research animal, the fruit fly. Not to mention the nematode. But anyway . . .

I didn't know what to say that I hadn't already conveyed: I grew up on a farm, eating—after killing, skinning, and cutting up—food that had a face. I wear leather. I regularly kill mice, mosquitoes, and ticks. Nothing seemed to work on this guy. But grudgingly he did agree to let me come, grumbling that since I was in Maine the least I could do was bring his wife, a displaced native of Quebec's Gaspé Peninsula, some lobsters.

I delivered the carton of sacrificial crustaceans in the darkening night before our interview, foraying into a North Carolina countryside that reminded me of home: The farther from town, the greater the number of vehicles elbowing into the soil around the houses, like herds of sleeping livestock. Then off in the woods was a glimmer of light and a pretty minimanse.

He met me at the door, a gentle soul with the sad eyelids I think of as a French feature. Now shy and formal, Caron led me into a kitchen where his wife beamed. Granddaughter Olivia, a dark-eyed sprite, uttered not a word in the ten minutes I stayed. Grandson Logan squared his shoulders and informed me that in his experience the mice at his grandfather's lab act quite a lot like people. Caron smiled.

He was going to be a tough nut to crack. Shy scientists are difficult to interview. And even shy ones are a lot easier than shy ones who fear you're going to set fire to their offices and fling fake blood on their families.

"Tell me how I know you're not PETA" are his welcoming words when I arrive at his office the next morning. He is pleading. He wants to trust me, but he can't shake his anxiety. I can relate to the anxiety, heaven knows, but I am at a loss. I have delivered unto this man ten living animals that I understood he would plunge, still alive, into boiling water. Even confirmed carnivores consider this savage. And still he suspects me. I wonder if there might be a pigeon handy whose neck I could wring.

"Call my editor," I beg. "Call the researchers I've already visited. Look at my website—in one photo from Mongolia I'm wearing a fur hat that could be from an endangered wolf. Please."

He writes down the editor's number. He sighs. "OK." That sad look is killing me.

Of course Caron is not going to let me suit up and go meet the mouse version of my impulsive, distractible husband in the lab. His anxious amygdala is still working the bullhorn: *She's a spy! She's a fruitcake! She'll set all the mice free!*

But he does open a video file. Two mice in a condo amble hither and yon, sniffing each other and washing their faces. And around this nucleus whirls a mouse electron. This black cutie skitters over their backs and under their stomachs and bounces off their faces. He's unstoppable. Around and around and around the cage he goes, leaping, rearing on his hind legs, tearing around some more.

"The made-for-TV movie," Caron announces. He's seen this footage a hundred times and still he's smiling. It's funny. It's cute. Anyone who knows a child with ADHD would smile in recognition. But only for a while.

This is Madcap Mouse's response to a new environment. All three mice have been put into an unfamiliar condo. The two normal mice also explore eagerly at first, but after about thirty minutes they settle in. Madcap Mouse is still bouncing off the walls after six hours.

"They'll do this until they croak," Caron says. A new environment is so stimulating to them that they'll forsake all else to explore it. Their accelerator pedal is working brilliantly—look at this, sniff that, climb here—but they have no brakes.

This mouse is a dopamine mutant. Created in 2001, it's now used in pharmaceutical labs to develop new drugs for children with Attention Deficit/Hyperactivity Disorder (ADHD).

Dopamine, like serotonin, facilitates communication between nerve cells in the brain. And like serotonin, when it's out of balance it causes your brain to malfunction. But whereas serotonin regulates your social interactions, dopamine oversees your motivation and physical activity. (In reality each does much more than that.) I think of dopamine as the spurs in your brain. Its role is to drive you toward behaviors that promote a social animal's fundamental goals: eating rich foods; mating; being kind to others; even punishing society's cheaters. The dopamine system makes you want to eat every day. (The same system can make you want heroin every day.)

But dopamine can also make you unhealthy, and unpopular. On a typical day, a typical person might experience the impulse to eat a quart of ice cream, yell at his children, run out of his office into the sunshine, and flirt with his neighbor. If your dopamine system is in good shape, you'll automatically inhibit those impulses. You won't have to think about it. If your dopamine system is out of whack, you won't. Tourette's syndrome, wherein people may blurt out profanities or impolite truths, is a classic dopamine disorder. Parkinson's and Huntington's, whose disordered physical impulses are on exhibit for all to see, also have roots in the dopamine system. But ADHD is the dopamine disorder most people are familiar with.

Just like serotonin, dopamine has its own fleet of street sweepers that collect and recycle molecules from the intersection between nerve cells. This keeps a steady supply of dopamine in the intersection, ready to unlock the gates so that a signal can continue through the brain.

In Madcap Mouse, Caron has sidelined the street sweepers. With only 10 percent of a normal fleet of sweepers carrying excess dopamine back to its home neuron, the chemical builds up in the intersections. Madcap Mouse has five times more of the stuff cluttering its synapses than a normal mouse. As a result Madcap initiates physical motions constantly, and he seems unable to inhibit them. The mouse's exploration instinct drives him to explore. And explore. And explore. The drive to eat is misdirected to some backwater of the brain. The message to drink is a dim murmur. The message to take a nap never gets through.

Any new situation brings on this frenzy, Caron says. Novelty, which is somewhat stimulating to all mice (and people), sends this mouse into orbit. This is not an animal who takes change in stride.

As he talks, Caron seems to be forgetting I'm an animal rights terrorist. Unfortunately he's now mistaking me for a cell biologist. He's talking a blue streak about chemical reactions, and to make sure I'm following along, he's scratching pell-mell with a pencil on a blue pad. Here's a sample:

$$VMAT2 = SAME\ NE,\ DA,\ 5HT.\ HISTAMINE$$
$$TYROSINE\ TH \rightarrow LDOPA \rightarrow AADC,\ DOPAMINE \rightarrow NE$$

What I want to know is how he determined that Madcap Mouse was a good model of a hyperactive human. How did he measure Madcap's impulsivity?

The answer goes back to the Lesch lab in Germany. There I was introduced to the "open field." It's a large box with no top. The floor is gridded, and the center lit from above. When you lower a normal mouse into the middle of the open field, she will scoot to the edge. Sticking to the safety of the walls, she'll explore the perimeter, making just a couple of forays back to the center to be sure she has overlooked nothing. But a mouse with high impulsivity will charge right in. The five-minute track of a normal mouse looks like an O with a line or two across the middle. The track of an impulsive mouse looks like a snarl of yarn.

I can't imagine how Madcap Mouse could stay out of the center of the open field even if he wanted to. He appears to have zero control over his own trajectory. He just goes. Fast. And indeed, his performance in the open field wasn't notable for his time spent in the center, but rather for his total hyperactivity.

Another measure of Madcap's personality came from a traditional maze that rewards learning with food. Normal mice were able to solve the maze in two and a half minutes. But Madcap Mouse was a terrible learner. Instead of putting his nose to the grindstone, he sniffed in corners. He reared to investigate the walls. He dithered. He was distractible. The Madcap Mouse bombed out of maze class, with most of the individuals unable to learn the route within the allotted five minutes.

The mouse's physical impulsivity was remarkable. But its social skills were also disordered. Madcap Mouse was a little tyrant. If you put four normal mice in a new cage, they'll scrap among themselves to see who's top mouse. It takes them about a day to sort out who's number one, two, three, and four. Then they can be buddies. But put four Madcap Mice in a condo and watch the fur fly. They battle for days. In the end the meanest bully rules the condo, and the other three skulk about with a chip on their shoulder. They're quick to take offense and resume scrapping.

This was reassuring, in a way, because it mirrored the social behavior of people with disorderly dopamine. Like Madcap Mice, many people with ADHD, Parkinson's, and Tourette's exhibit a tendency to bite. (I mean that figuratively.) Aggression and combativeness are common features of those disorders.

Of course, one of the best tests of whether a mouse is acting like a human is to give it the same drug you'd give a human. One peculiarity of ADHD is that amphetamines—drugs that cause most people to speed up—actually allow people with ADHD to slow down, to control their impulses. And sure enough, the amphetamine Ritalin allowed the mice to decelerate, just as it does hyperactive children.

When Caron is ready to be rid of me, he escorts me into the basement. Not quite to the mice. But to the mouse whisperer.

"Ramona Rodriguiz knows how mice think," he says as he hands me off. I'm actually glad to change. I've noticed that most researchers have no idea what a real mouse acts like, out in nature. Ramona Rodriguiz isn't like that. Her reputation is for knowing mice so well that she spots tiny differences in behavior. Tiny and very crucial differences.

She has soft hair, a soft sweater, a soft and pleasing voice; she even has a plush animal on her windowsill. Her minty office is a haven of femaleness in the beige wastelands. Yes, her desk strains under piles of research papers, but these are tidy piles. Her bookshelf includes the one, the only ever, the fascinating book about how real wild mice behave. She puts her hand on it almost without looking. She found it on eBay. When I squeal with excitement, she just as deftly retrieves a photocopy that I can take with me.

"Oh, the dopamine transporter knockout mice," she says as if remembering a favorite student. "We trained them to press a bar for food. Well, they start pressing for food—and they don't stop. Food will come down the chute, and they won't stop to eat it. Or they push the bar and make two laps around the chamber. Or—they're small—they'll climb right up in the chute. You open the test chamber to see why nothing is happening, and you see a little tail disappearing up the chute."

She talks fast, and never pauses, but every single thing she says is interesting. Regarding Madcap Mouse's performance in a maze, she suspects they suffer not from a shortage of smarts, but from an overload of options. "In a maze, they show you head pokes, stretches, bobs—mice are talking to you all the time. Almost all mice will give you clues if you watch them closely enough. We realized that maybe these were having trouble making up their minds. It does look like they have trouble making decisions."

This is something I've seen in one of my favorite people with ADHD. The simplest decision can bring on a state of agonizing paralysis. In front of a pizza display he'll waver between ham and pepperoni, pleading, pleading: "Which one would I like better? Which one? Which one would I like?" I'll capitulate and choose for him. Halfway to the counter he'll look stricken. "I want the other one!" It's not a game. To all appearances he feels tortured.

"It's so amazing, the similarities between children and mice," Rodriguiz agrees. "My PhD is in child development. You watch these animals and say, 'Gosh, I used to see this on the playground.' "

Caron pokes his head in, on his way home for the day. Nodding toward Rodriguiz, he asks me, "So was I right?"

"Definitely clone-worthy," I answer. "I want one."

"We all do," he says.

I'm starting to worry about catching my flight, and she jumps up to show me the way out of the building. I notice the dog trinket hanging from her key chain. "Aww," I say. (It's a dog-person thing.) "Yeah," she coos. And the story of the dog spills out. I could stay for days. I have to go. "You can't come home with me?"

She gives me her card, adding her home address and three phone numbers. And returns to her mice.

Dopamine can be seen as a "go forward" chemical. An animal whose dopamine system is maladjusted in some way may not want to embrace life. Rats drained of dopamine won't eat. Mice with a dopamine shortage show little curiosity about new objects placed in their cage. And a human with a mild disruption of dopamine, far milder than Madcap Mouse's, seems to crave outside stimulation to raise his dopamine activity to a comfortable level.

The low-dopamine personality doesn't get the thrill out of his dopamine that other people get. His brain feels a nagging dissatisfaction, a constant craving. He yearns for stimulation, for something new. He talks to strangers. He eats impulsively. Or shops compulsively. When a pretty red car appears with a "For Sale" sign on it, he pulls out his credit card. The dopamine flows and—ahhhh—he feels better.

Like Madcap Mouse, the human may be both physically and socially impulsive. Sometimes they're both at the same time, if their love lives are any indication. Extraverts are apparently more likely to act on their romantic impulses than more introverted types. An impulsive man is more likely than most to cheat on his spouse; an impulsive female is more likely to skip out on her spouse altogether. Extraversion is the strongest predictor of which people will have the greatest number of romances over the course of their lives.

Obviously, impulsivity carries risks. Extraverts have an elevated risk of being hospitalized for accidents than others do. They also take more risks with alcohol, drugs, and sex. And their impulses turn easily to compulsions that drive them forward with such insistence that it can kill them. Addiction to alcohol, drugs, gambling, exercise, or anything else that dopamine pushes animals toward, is common in Extraverts. And impulsivity is also a risk factor for suicide.

The extreme version of the impulsive personality would seem to be ADHD. As my young friend's amphetamines wear off at the end of a day and the dopamine supply in his synapses drops, his impulsive personality shines through. It blazes through. It sets the candles alight, and then the curtains. Everything gets his attention—my haircut, the dog's new collar, the pot on the stove. In his yearning for the dopamine high of social interaction he converts every observation to a question, and out they tumble in a flood: *When did you? Where did she? Why is that?* The stimulation of crashing onto the sofa is irresistible: He launches himself. The sensation of twirling calls to him: He spins until he falls, laughing wildly. I ask him what it feels like.

"I feel wild inside," he says, beating his heels on a chair rung, frustrated. "Like I don't know what to do. I just feel wild inside."

If you'd like to test drive that sensation, take amphetamines. On the street you'll want to ask for "speed." I understand you can also procure it in schools, from teens with an ADHD diagnosis. Cocaine is another option, but some personality-altering chemicals are more illegal than others, and this is one of them. Both drugs "work" by increasing the amount of dopamine in the crossroads between your nerve cells. From personal research in my teens I recall these drugs produced a percolating readiness for action. The percolating sensation was physical as well as psychological. It was like an adrenaline rush, but without any fear. It was a restlessness that had nowhere to go.

Too much of these drugs will tilt you into a hazardous state of psychosis. Yes, you'll be wide awake and alert to all the stimulation the world has to offer. But you'll also be subject to euphoria, delusions, and hallucinations. If you're mistaken for a schizophrenic, it's an honest mistake: Schizophrenics are thought to suffer from a surplus of dopamine in some of their intersections.

To experience a low-dopamine state, you would take the dopamine-reducing drugs schizophrenics use. These can make you sluggish, dizzy, weak, spastic—basically give you Parkinson's disease.

This is exactly how scientists used to test drugs on mice, Caron had told me: by giving them other drugs that disordered them temporarily. They'd give a normal mouse one drug to make it act hyperactive and then, rushing to complete their experiment before that effect wore off, they'd try an antidote to calm the mouse down. It's a lot simpler, for both mouse and researcher, to start with a mouse who's naturally wild inside.

Impulsive Human

Somewhere in my notes on Madcap Mouse I scribbled "Spouse Mouse!" Husband John is an extreme Extravert, and extremely impulsive. It's this facet of Extraversion that makes me think ADHD is just another stop along the personality spectrum. It's an outward orientation, a reaching into the world, a hunting mode in which the attention just keeps roving on.

John is impulsive. He's not disordered. If you put him in an unfamiliar condo he won't run to exhaustion. But let's just say that if money were no object, he would adopt a stray BMW every week. Let's just say his prefrontal cortex (PFC) is not an iron fist of rationality clenched on his personality.

The PFC is the bulge of brain that stuffs your forehead. Other mammals have a PFC, and it apparently does the same job for them: The PFC is the CEO. When the brain encounters a stimulus—a cute old BMW on a car lot, just to pick one completely random example—a network of brain departments perks up. Each forwards a memo to the PFC:

Sensory Cortex: *Red! Shiny!*

Motor Region: *Let's get our hands on that wheel!*

Nucleus Accumbens: *Pleasure, pleasure, pleasure!*

The PFC considers this input. It also weighs its own experience: *I have two already. Children are starving in Haiti. Hannah would break my kneecaps.*

The PFC weighs the whole bundle of data, and then my dear husband buys the BMW.

I have a PFC of steel. I drive the *Consumer Reports* "best small car" and I don't care what color it is. I eat my favorite part of dessert last. I am monotonously moral. As we'll see when we reach the Conscientious factor, I am the least impulsive person on the planet. To walk a mile in my PFC, just do everything the surgeon general, the U.S. Constitution, and the Dalai Lama advise.

To walk a mile in my spouse's PFC is a lot more amusing. To visit his world, have a drink. Have a couple. Alcohol relaxes the PFC, allowing the brain's more impulsive regions to stretch their legs: You sing without shame. You dance without rhythm. You buy cars without compunction.

Normally, the PFC is your prioritizer. From the bewilderment of stimuli in your environment, it helps you to focus on one thing—the book you're reading—and to fend off distractions. A dog barks, a catbird mews, the radio plays a new song, but your PFC holds your attention steady on your book.

That is, unless you're more open to the world's distractions. My spouse's PFC is one of those CEOs who manages by empowering. Rather than rule from above, his PFC delegates. In this democratic environment, every idea gets equal treatment. So this is what happens when we step out the door to talk about what color bricks to buy for the patio: His eye falls on a loose piece of siding on the garage. "I need to fix that." We walk toward the backyard. "Let's put lilies along the fence!" he proposes. "Should we cut that tree down? I want to build a grape arbor there. Should I use cedar or pine?"

My PFC remains by the door. It came out here to talk about bricks. It will not be diverted.

It's true that everybody's PFC has a special mandate to take note of new conditions. All animals share this bias for novelty. After all, any change in the environment could signal new opportunities or new threats. So when you scan the landscape—even inside your house—your attention subconsciously takes note of what's new.

It's my contention that a more impulsive PFC goes beyond subconscious note-taking. This is definitely true of ADHD children, whose eyes buzz over their environment like workaholic bees. From what I see, even a brain that is in the normal range of high impulsivity becomes fully aware of each novelty in its environment. My spouse is the first male I've ever known who notices when I trim a quarter-inch of hair off my bangs, let alone get a full haircut. Nothing gets by him. There is no hiding the candy, no stealthily switching off the television when he's in the other room. "I'm listening to that," comes a faint voice from the other end of the house.

But he's not superhuman. With all those subjects and objects hurtling around in his head, things do get lost. He lets the dog out but neglects to shut the gate. He leaves his office window open, which sets off the burglar alarm in the night. He puts burgers on the grill and then goes around front to pull a few weeds . . . And that was all yesterday.

Research on personalities like John's—with high, but not officially disordered impulsivity—is rare. If there's nothing wrong with a personality, why waste money studying it? So to understand a PFC that's in the normal range of impulsivity we're left to extrapolate from the extremely impulsive PFCs found in the heads of people with ADHD.

The "stop signal" is a classic way to measure human impulsivity. You sit at a computer, and whenever an X appears on the screen you click a button—quickly, before the X vanishes! But not too quickly, because if an O appears a split second after the X appears, you're not to click. It's demonic. The X appears, your dopamine starts to pound your finger against the button, and the infernal O appears. Can you stop in time? Kids with ADHD can't, not nearly as often as other people. And inside their heads as they struggle, their PFCs aren't as active as other people's either. The CEO is having a hard time reining in the departments.

Another means to torture your impulsive friends and family is to teach them one response really well: Whenever you see a car with one headlight the first person to yell "padiddle!" is the winner. Then change the rules. Now when you see a car with one headlight the first person to take a deep, cleansing breath wins. If you could see inside the heads of your impulsive loved ones as they continue to yell "padiddle" for months, you would again see a PFC that appeared halfhearted about implementing the new policy.

MRI research has shown that the more easily—impulsively—a child shifts his attention from one thing to the next, the more laid back his PFC is during the experiment. Even the siblings of children with ADHD may have quieter PFCs, according to some research. Presumably these children are impulsive, but not to such a degree that they've been carted to a psychiatrist for treatment.

As CEOs go, the impulsive person's PFC takes a hands-off approach to management. This PFC does not run a tight ship. Impulsive brains are more like Viking longboats where the wine has run too freely, the oars are all banging into one another, and the star chart has blown overboard. It makes for an unpredictable voyage, but that's part of the charm.

The onrushing attention of an impulsive person can make the people around him feel, well, unarresting. But recognizing the biological basis for my spouse's behavior has improved my marital satisfaction. Instead of getting mad when he asks half a question then leaves the room, I now continue with my life until or unless he locates and kills the mammoth in the living room, and returns.

Evolution of Impulsiveness

What are the benefits of an impulsive personality and a fleeting attention span? The risks, as we've seen, are substantial. More accidents, addiction, risky drug use, risky sex, and suicide are among them. Moreover, in the current culture, where people are expected to work doggedly on one task at a time, an impulsive personality can make you a target for both laughter and frustration.

So what's it good for?

The impulsive novelty seekers in my life are explorers. They're stimulated by novelty in its own right, not by the chance of making a thrilling discovery. They notice everything, and they remember most of it, too. They're hypervigilant like we-the-Neurotic, but they don't scan only for danger. They're actually more attuned to the opportunities and rewards that often accompany a change in the environment. In a new condo or a new country, an impulsive person comes to life, exploring and taking note.

In impulsivity we find the opposite of the retreating personality that acts on anxiety and fear. And if you recall the saga of Mitzi and Maxi Mouse, you'll remember that low-anxiety Maxi Mouse fared well in an environment where the harsh conditions encouraged a mouse to take some risks. Impulsive behavior can serve a human in the same way.

Imagine you're on an airplane that crashes in the jungle (no injuries, of course). Who do you want to hang out with? The Neurotics who stay near the familiar airplane licking salt out of the empty pretzel bags? Or the impulsives who venture forth, picking strange fruits and taking that first bite? Do you want to hitch your fate to the guy who has the discipline to start a fire by grinding one damp stick against a damp piece of wood? Or would you rather follow the one who seems to notice every creak, snuffle, and snort in the forest?

It would be ideal if you didn't have to choose. If the divergent personalities all cooperated, they could pool their strengths to keep the whole population alive and well fed. And that is exactly what evolution has provided. In some parts of the United States one in five children is labeled with ADHD. Although many outgrow the diagnosis, others retain their impulsive natures into adulthood. This is certainly a case of overeager diagnosis, but the statistic hints that a large portion of people are naturally leapers, not lookers.

I have a hard time seeing ADHD as a disorder. I think of it more as a *difforder*, as in different. I think ADHD is exactly the personality you'd want if you hunted for a living. These personalities have high energy. They notice everything in their environment. Many people with ADHD also have the ability to "hyperfocus," in which the entirety of their awareness comes to bear on a single task. That kind of feature is nice when you're stalking a deer and don't want to swat the mosquitoes that bite you.

Perhaps evolution will trim their numbers in the future. The pace of life in the Westernized world is very different from the pace just four generations ago. On a small farm, with no radio or television, even a person whose attention was drawn to *every single thing* was probably not stimulated to the point of exhaustion every day. Where traffic moved at the pace of horses, accidents were less common and less deadly. (Outside of the factories, anyway.) Today a blizzard of stimuli threatens to clog even the most focused person's attention. If your mind should stray from the task of walking a city street, you'll be flattened by a hurtling chunk of metal. And temptresses of addiction beckon to impulsive people from every corner: One in five gambling addicts has ADHD; half of all adults with ADHD in a German survey reported a substance addiction.

So perhaps the impulsive personality is on the wane. I hope not. Furthermore, I have reason to hope: Recall that the impulsives have a high total number of romantic partners. Thus they may also have a high number of offspring. And evolution smiles on the personality of those who procreate.

Extraversion Facet:

ACTIVENESS

ACTIVENESS INDICATORS	RARELY	SOMETIMES	OFTEN
I'M UP FOR SEEING FRIENDS	❏	❏	❏
I HAVE TOO MUCH TO DO	❏	❏	❏
I'M NOT GOOD AT RELAXING	❏	❏	❏

THIS GIVES YOU A QUICK look at where you land on this facet. If your answers tend toward the "often" side, you're higher in that facet.

The active personality is the one that will pencil you in for next Friday because she has something going on every night until then. She's the one who's working on seventeen projects and doesn't mind starting another one if you need a hand. She walks fast, talks fast, and it's not clear whether she sleeps.

If you ask her to take the bus to work, sit in an office all day, go home to microwave a meal and relax on the couch . . . she won't relax. Her knee will jiggle. She'll review her mental list of chores a hundred times. Her eyes will wander the room, seeking a door she can sneak out of.

The low-active personality, on the other hand, will not opt to bicycle to the office, work like a man possessed, bike home, take the kids for a hike in the hills, and make dinner from scratch before doing a few loads of laundry and bathing the dog. People who score low on the active scale aren't necessarily depressive or Neurotic. Some people are perfectly cheerful and outward oriented, but content to sit and watch.

Active Mouse

Hyperactivity is a common side effect of messing with a mouse. Scientists such as Marc Caron have created dozens of hyperactive mice, some by intention and some by accident. "Hypoactive" mice, who lie on the couch clutching the remote and squeaking for beer, also turn up by accident but they're less common. Even within the dopamine domain, which contains many genes, some gene alterations will produce busybody mice while others produce loafers.

Beyond the dopamine system, mice go speedy in response to all manner of genetic tinkering. Altering hormones, metabolism, chemicals that help genes to express themselves, and even growth factors can bring out the rowdy in a rodent. In fact, about sixty different genes, when toyed with, have produced hyperactivity as a main effect or side effect.

This raises the possibility that a high rate of activity is the default condition of mice—and perhaps man. That would suggest, in turn, that earlier, more primitive animals scurried around at a faster pace, as plankton do. Perhaps evolution applied some brakes to the high activity when more complex animals faced complex challenges. Busyness prevents you from stalking prey patiently; from freezing when a predator appears; from resting when you may, so that you can run when you must.

So hyperactivity may be a fairly natural condition for an animal. There are certainly worse fates that genes can impose on a creature. Ceaseless scurrying does potentially expose an animal to more predators, and demands a higher rate of food input. Regardless of the drawbacks, it's a strategy employed by animals still living today. The short-tailed shrew right here in my neighborhood hunts day and night, eating its way from insect to worm to baby mouse. This lifestyle does expose shrews to cats, foxes, and owls, and the animal's metabolism is so high it can starve to death in a matter of hours. But no matter. Shrews are probably more numerous than mice here in the suburbs, though few people ever see one. Hyperactivity is a perfectly functional lifestyle. But it's not for every species. It's hard on a human. And mice.

Testing a mouse for hyperactivity is refreshingly simple. There's no tail-hanging. There's no elevated plus maze. The mice just scamper around.

In one method you count "rearings" as your mouse scampers. Mitzi, my anxious mouse, does this in any new situation, including her afternoon rambles across my desk. At any wall or barrier she lifts her paws, stretches her nose up, and whiskers the air for all she's worth. It's an indication, researchers think, of a genetic drive to explore. If you already have an open-field apparatus for testing mouse anxiety, you can use that as your activity apparatus. During the five minutes your mouse spends in the open-field box, you count her rearings to see how she compares to other mice. Infrequent rearers are couch potatoes. Eager rearers are exuberant explorers.

For a more automated method, hang a video camera over the checkerboard of the open field, and program your computer to translate the camera's data. Lower your mouse onto the checkerboard and let 'er rip. You'll get an automatic tally of how many centimeters your mouse travels as she explores the apparatus.

Testing mice for activity is not much different from watching a dozen young children arrive at a new playground. Some will settle in the sandbox and push bulldozers. Others will run, and rear, and climb, and rear some more. Some are not energized by exploration. Others find it exhilarating. Some seem inactive until their anxiety wanes and their energy bursts forth.

Testing mice for the genes that underlie hyperactivity is more difficult. Scientists have combed the dopamine system—its intersections, its street sweepers, its raw materials—hunting for one powerful gene that makes a big difference. No such luck. They've found many genes that matter, but each makes only a small contribution to a finished personality. As with serotonin's convoluted relationship with depression, dopamine has a complicated relationship with an animal's energy level.

The reason so many scientists have created hyperactive mice is, of course, to improve treatment for ADHD. For most people the active facet of Extraversion isn't noteworthy. Humans come in such a variety of activity levels that few of us look out of place. The major exception is the hyperactive contingent, those students who are bouncing off the walls when it's time to sit down and read. You don't have to put these kids in a box and count their rearings to determine their activity level. You can just ask their teachers and parents which kids are squirmy, fidgety, climbing on furniture, running, talking nonstop, talking loudly.

It is only since the 1980s that such kids have been treated as though they have a chemical imbalance. Before that, they were treated as though they were a pain in the butt. Lots of them dropped out of school. Others discovered on their own how to manipulate their dopamine system for a more comfortable existence.

Marc Caron's sad-shaped eyes had looked genuinely sad when he recalled hearing from one of these people a few years ago. "I got a letter from someone in prison. He told me about how his friends were into taking coke. He was really hyper already. And they said, 'Look what it does to us. You're going to explode!' Well, he found it made him feel normal. He went on to become an addict. And a criminal. And he wanted to be a subject for our research."

A lot of people with superactive personalities slip past the teachers and doctors even now. And they continue to stumble onto their own versions of a cure. The "self-medication hypothesis" proposes people whose activity level is uncomfortably wild find their own calming drugs in cigarettes, cocaine, and amphetamines, all of which adjust the brain's dopamine levels.

Active Human

Except for ADHD, the activity level of humans goes unstudied. A personality that scores low on activity is no handicap. Among my friends who measured their personalities for me, my friend Leo came out on the . . . bottom? The top of the low-activity heap? How to say this without sounding judgmental?

Our culture is so in love with the overachiever that we have a hard time seeing the virtue in downtime. In fact, so restless are all my friends that Leo scored lowest in activity even though he routinely works a fifty-hour week. He puts in a full week consulting for businesses around the country, then spends many evenings in his home office. Perhaps it's easier to see who's high-active and who's low-active in a culture that's not so bent on being busy.

Or perhaps it shows up in the privacy of the home. In the privacy of Leo's home his wife calls him her "sleep coach," the man who can nap anywhere, anytime. Before they met he ate out at every meal rather than rattle the pots and pans in his kitchen. And at their engagement party someone gave him a leather tool apron as a gag gift. Leo does the obligatories—work, walk the dog, work out—and then he takes to the

ACTIVENESS | 71

couch. Personalities like that of my spouse, who last night came home from work and then toiled at building a new fence until dark, baffle him. "Why?" Leo asked when he called. "Don't you own a phone book? There are people who will do that for you."

There's not one harmful thing about being a low-activity personality. Leo is beloved by all who meet him. Nobody cares in the least that he doesn't spend his weekends gardening or installing a brick patio. (Well, his wife has confessed she'd like to rent a "Holiday Husband" for two weeks each year, one who would get her a gift, accompany her on a sleigh ride, and such. But in Leo's defense he is nominally Jewish.)

At the other end of the spectrum a high-active personality is also harmless. Busy beavers get a lot done. Because they're often Extraverts they're usually willing to spend their energies helping others. Also because they're often Extraverts they don't usually stress out over their long "to do" list. And the activity itself certainly contributes to the long life span of Extraverts. These blessed beings do enjoy longer, healthier lives than average. In part, that's probably due to their low levels of stress chemicals. But as a group, Extraverts are more physically active, and that conveys a health benefit.

Now, being married to an active human myself, I have become aware of a few drawbacks. The one I live with is very active, but also showcases the impulsive facet of Extraversion. As a result, prioritizing is not his strong suit. He might start the day cleaning the basement, but discover a hole in the foundation that would be ideal for running a water pipe to the back of the yard where it would be nice to have a hose. At the end of the day the basement is a pit of chaos and there's a trench across the lawn. Last week his energy drove him toward removing the transmission from his pretty red car. But a week later he began the aforementioned fence construction. The transmission remains in the garage, the car up on blocks. And already he has mentioned a dozen contenders for his next undertaking. I haven't heard lately about the pantry he was going to build me last month.

That combination—many projects and no priorities—also results in a fair amount of lateness. That can be tiresome. And as the years add up, the hard-driving personality may find that its natural buoyancy is periodically weighted with those aches and pains so familiar to the weekend warrior. The dopamine is willing but the flesh is weak.

Which may actually be the reason some personalities are underpowered with dopamine: to drive an animal forward even in hard times. One of the dopamine genes, DRD4, may have evolved to do just that—to ensure that a few humans will always be pushing the envelope, expanding the territory. In my favorite study of DRD4, researchers went poking around various populations to see who might be benefiting from what you could call "Restless Human syndrome."

If dopamine motivates an animal to act impulsively and to seek food, reproductive opportunities, and novelty, what modern-day group of people would you expect to carry dopamine genes that are a bit hyperactive? Where might you find a bunch of people with a certain "long" version of the DRD4 gene that's gaining a reputation for lurking in people who are superactive and love novelty? How about nomads?

Unfortunately, few nomadic people remain on the planet. Most of us have settled down. So a team of curious scientists collected a batch of DRD4 studies conducted on thirty-nine populations around the world and searched them for interesting patterns. They used the evolution of languages to estimate how fast each language group had moved across the planet to its current location—from Asia to the Americas, or from West Africa to South Africa, and so on.

The farther a population had drifted in the past thirty thousand years, the richer it was in that long version of DRD4. So in homebody China virtually no Han people have the long version of DRD4. But, to trace one path of migrating people, from China let's cross the Bering Strait to Alaska, then drop south through Central America to the jungles of Colombia: In the Ticuna people of the western Amazon, four out of five people carry the long DRD4. The same pattern emerged in each case. The farther people had migrated, the more common the long version of DRD4 was among them. This is the restless version of the gene. This is the version that gets bored with rebuilding the transmission and builds a fence.

Curious about what effect the restless version has today, a second team focused on one tribe of people in Kenya. Some tribe members had settled, others were still migratory. All were undernourished, finding just enough food to get by. But the nomads with the long version of DRD4 had an edge. By a substantial margin they had a higher body mass index. The settled people with the same version of the gene got no benefit. In fact they were much skinnier than settled people without a long DRD4 gene.

How does it work? Scientists don't know. It may be that a hunting-

and-gathering lifestyle profits from impulsivity and high activity, whereas those features are a liability for settled farmers with lots of neighbors. It could be that the low-dopamine status of the nomads drives them to work harder at foraging for food, an option that isn't available to farmers.

A sidebar to this story of wanderlust is that the same long version of the dopamine gene may affect how attached a child is to his parents. To phrase it another way, it could affect how willing a child is to strike out on his own. Some children show "disorganized" attachment right from birth. They just don't orbit around their parents the way most infants do. If their parents are nurturing and attentive, these children can develop normally. But if their parents amount to a stressful environment, the children may migrate right out of the nuclear family.

Incidentally, DRD4 is also among the genes associated with ADHD. It would be interesting to know whether the superactive and impulsive personality we call ADHD is also more common in nomadic and far-flung populations.

Is your personality too active? Not active enough? For most people, culture barges in and answers that question. My culture likes to pretend the national personality is one of pioneering, sod-busting, gold-mining, skyscraping adventurers. We admire people who always have eight things going on. On the other hand, our cultural reality is one of extremely low physical activity. As that lifestyle gains followers, acceptance for the low-active personality is growing, too. The range of personalities we embrace as "normal" widens.

And that's a good thing. A personality can be called disordered or diseased only if it causes, you know, dis-ease. And that dis-ease is often caused by your culture, not your biology. The same personality that causes friction in a schoolroom might, in a different culture, cause you to be an excellent hunter and explorer.

Evolution of Activeness

An animal's activity level will determine how quickly she explores and exploits her environment. But it also burns up energy, energy that will have to be restored with fresh food supplies. Humans have a mouthful of proverbs to describe the superiority of each strategy.

"Make hay while the sun shines," urge the active ones.

"Stop and smell the roses," admonish the energy savers.

"Strike while the iron's hot!"

"Slow and steady wins the race."

Humans confront a vast number of challenges to our energy budget every day. We make decisions about how to transport ourselves to work, how hard to work, whether to change jobs, how to amuse offspring, how to amuse ourselves, what to feed ourselves, how to help strangers and friends and family. The options are many and each requires a certain expenditure of energy.

Some humans fall on the conserving end of the spectrum, holding back their vigor. Others will invent new ways to spend energy if there's no work to be done. *I'm going out to shoot hoops! Wanna play chess? Let's go to Colombia!*

Each style has merits. When opportunity knocks, perhaps in the form of a herd of juicy deer, a naturally conserving personality will be ready to take advantage. He is well rested, he has no broken bones, and he can now jump up and pursue a great meal. The naturally active hunter may not be so ideally positioned. She may have spent the previous evening playing kick-the-hedgehog with the kids, and is now laid up with a hedgehog spine in her toe.

The active personality is going to shine in a new environment. Fired up by all the novelty, she'll explore nonstop. She'll be the first to discover a spring with clean water, a tree with ripe fruit, a watering hole with deer tracks pocking the mud. And yes, she'll also be first to experience the stinging ants that live in this environment. And the pool of quicksand. And the hungry hyenas shadowing the fruit tree . . .

So here's my dime-store thinking on hyperactivity, the type we think of as a disorder. It is, in part, the response of a developing fetus to signals that reach it in the womb, warning of tough times ahead. Consider:

ADHD has a strong genetic component, like any personality feature. But a person's activity level is also subject to environmental conditions that can ease that level down, or jack it up. The environment might even jack it high enough to qualify as hyperactivity.

Scientists are particularly interested in how ADHD relates to the environment of the womb. For instance, a pregnant woman's tobacco and alcohol use both alter that environment. Stress impacts it: Women who are more troubled during pregnancy tend to have children with more pronounced ADHD behaviors. And the admittedly vague "low socioeconomic status" may also be a wrench that fine-tunes a developing brain's activity setting.

What I wonder is if the "jack up the activity" signal can take a hundred different forms: as toxins in the mother's blood (nicotine, alcohol, other drugs); or as stress hormones that reach the womb; even in the form of a mother's low fat level (suggesting famine) or the opposite (fat cells cause inflammation that affects an entire body). One or many such signals might serve as a chemical update to the fetus, allowing him to adapt even before he's born: *Dangerous environment out here, buddy! It's getting the best of your mother! Prepare to face an unpredictable world!*

And a fetus who already leans genetically toward high activity receives these chemical warnings directly in his growing brain. He develops a dopamine system that's even more active than the one his genes sketched out.

The end result is a personality that's primed to thrive in a chaotic environment. This little person doesn't mull over opportunities. He grabs them before they disappear. He doesn't lounge around, sniffing the flowers. He roams, gathering information about this unreliable world. He doesn't lose himself in contemplation. He keeps an ear to the ground, an eye on the horizon, a finger to the wind. He takes it all in.

Even the notion that a child can be genetically primed to detach from inadequate parents supports my pet hypothesis. What an excellent feature, to walk away from a lousy home without looking back in guilt and regret! To seek your fortune beyond the horizon, unfettered by family ties and apron strings!

I take further encouragement from the high rate of hyperactivity in gene-altered mice. When a gene alteration does change a mouse's activity level as a side effect, that mouse's activity level is twice as likely to go up than down.

One implication could be that hyperactivity isn't much of a hardship. If it were, it would be buried deeper, so to speak, and wouldn't pop up so frequently. Hypoactivity, by contrast, is harder to bring out, and perhaps a more risky personality type. The risks aren't obvious in laboratory mice. For them, food always appears, and predators never do. But for a wild animal, be it mouse or man, a low activity level can be risky.

If you're a low-energy mouse, no other mouse is going to hunt and gather for you, or carry your babies to safety if a snake approaches. A low-energy mouse may expose herself to predators less often, because she spends less time wandering around the environment. But ultimately, she needs to perform her chores thoroughly if her offspring are to survive.

Humans can get away with a bit more loafing because we're so manipulative and charming. If you can tell a thrilling story, or always lend a sympathetic ear, perhaps your social services will compensate for your failure to contribute fruit, meat, water. But ultimately humans are social animals. We work together, and we keep score. We know who's trying to hoard a little extra energy for themselves. So perhaps in humans, as in mice, we're more likely to go hyper than hypo when our genes and our environment conspire to set our activity level.

Extraversion Facet:

CHEERFULNESS

CHEERFULNESS INDICATORS	RARELY	SOMETIMES	OFTEN
LIFE IS GOOD	❏	❏	❏
EVERYTHING WORKS OUT IN THE END	❏	❏	❏
I LIKE PEOPLE	❏	❏	❏

THIS GIVES YOU A QUICK look at where you land on this facet. If your answers tend toward the "often" side, you're higher in that facet.

The cheerful personality is optimistic and resilient. That doesn't mean that a person high in cheerfulness doesn't ever feel crabby or discouraged. But the cheerful personality is able to bounce back from difficulties and regain her composure. Cheerful people are easy to identify: They wear their emotions on their face.

A personality with low cheerfulness could be mistaken for a Neurotic personality. These people are more pessimistic and reserved, and are not prone to taking leaps of faith.

Cheerful Mouse

The Black 6 mouse may have won the dubious distinction of "most popular lab mouse" due in part to its cheerful temperament. It's one of the oldest research mice, bred a century ago by one Abbie Lathrop, a spinster who was breeding mice for a booming pet market. I don't mean the pet-snake market. People at the time were into pet mice.

Biologists at the time were into genetics. And when Lathrop sent some sick mice to a researcher who declared them to have cancer, the two trends collided. Geneticists recognized mice as a fast-breeding test animal for their research. Soon Lathrop was shipping hundreds of mice at a time to genetics laboratories. The rest is history. It's Black 6's history. Formally known as C57Bl/6, the black lab mouse was inbred from Lathrop's #57 mouse.

Is it an accident that this mouse became the most beloved mouse of all sciencedom? Or is it a question of personality? Perhaps I am biased. I married Black 6.

Black 6 is famously resilient. This mouse is about as cheerful as a mouse can be. Recall the stressful tests that bring out the anxiety and depression in a mouse. In the tail hanging test the "emotional" mice quickly cease struggling and appear to give up. In a beaker of water they float instead of swimming, again a sign of hopelessness. Not the Black 6. The Black 6 soldiers on. Even if you subject this mouse to chronic stress—you might tilt its condo for a few weeks—Black 6 shakes it off.

This mouse is socially resilient, too—in many ways, it's a little Extravert. When researchers double the number of female mice in a condo, most mice stress out. Crowding is hard on most creatures. Their immune systems sag, they eat too much, they're more anxious, and their stress chemicals rise. Black 6? Nah. (If you doubled the number of male mice in a condo, there would now be two males, which inevitably leads to fighting.)

I suppose you could even call Black 6 mice assertive. For one thing, they bite. They don't take kindly to human hands reaching into their territory and lifting them out. Chomp! These aren't cuddlers like the white mice. They have to be handled by the tail. And like my Extraverted spouse, they take competition seriously. Black 6 keeps strict hierarchies. The dominant female is likely to "barber" her subjects, clipping off little patches of their fur or whiskers. Mature males can't live together at all. The clear lines of power may contribute to the animal's peace of mind.

The assertiveness of Black 6 constitutes its only drawback. I remember this from the Lesch lab in Germany. A half dozen males of the white BalbC strain could share a condo without friction. They could even cooperate to arrange paper shreds in an elegant igloo in one corner. And when I walked through their room the little white faces pressed to the plastic in friendly curiosity. *Check it out! There's a new character on the People Channel!*

By contrast in the opposite bank of condos, occupied by Black 6 mice, scuffling was frequent and squeaking periodic. These were young males still able to tolerate one another. But only by the skin of their teeth. The battle for dominance was heating up. And their igloos! Disgraceful! Two-thirds of an igloo would occupy one corner, connected by trailing shreds to one-third of an igloo in a second corner. These mice had little interest in the People Channel. They would rear to sample the air when we came in, but they didn't crowd around to gawk, the way BalbC mice did.

But—and this is challenging for a Neurotic person to remember—the appearance of conflict and chaos does not mean the animal involved is experiencing emotional discomfort. Not in a mouse, not in a man. The state of arguing and domestic squalor that would make a Neurotic mouse quiver in the corner means little to a cheerful mouse. Inside those little black heads, stress chemicals were low. Anxiety was low. And the immune systems were ticking like Timex watches. Even during arguments, these were healthy, happy mice.

And unlike most strains of research mouse, these personalities were not tailor made. If they were consciously chosen as a good research strain it was probably from a strictly economic perspective: They survived the stress of shipping and handling better than others. Abbie Lathrop and the geneticists who patronized her mouse farm had no idea they were breeding a line of rugged, resilient Extraverts.

Cheerful Human

It's a blessed state for humans, too. I often envy my spouse's ability to shake off life's conflicts and cruelties. Through the luck of the draw, he got DNA that tunes out the facts that the planet is doomed and life is a ghastly struggle.

In truth, my cheerful spouse will sometimes arrive home at the end of the day with his shoulders drooping and his mood dragging. Recently one of his employees filed a claim of "repetitive motion injury" after precisely three days on the job. It's a pharmacy. The work isn't especially injurious. The claim was quickly denied. But dealing with a personality that specializes in deceit is tiring even for a cheerful Extravert.

On those days he'll flop onto the couch with a sigh. He'll spin out the tale, with more sighing and growling. He'll run his hand through his hair. He'll shake his head. He'll sigh once more. He'll turn on the basketball game and never think of it again.

(For contrast, me: I would drive home in tears, deeply wounded by this employee's betrayal of my trust. I would ruminate on what it implied about the human condition, how the species really was reprehensible in every way, how we deserve to be eaten by polar bears. The evening I would sniffle away in moody silence. In bed I would obsess over what I might have done differently, how I might have turned that employee into a good person if I had only tried harder, perhaps taken her on as my personal project.)

How does a cheerful personality retain its altitude in such a dismal world? Well, it's a lot like the way an anxious personality retains anxiety in such a promising world: attentional bias.

If you walk through the world noticing every fire, flood, and famine, you, too, can depress your brain. A diet of sadness, anger, fear, and assorted trauma will take its toll.

But the inverse is true as well. And cheerful personalities excel at noticing cheery information. It's a self-reinforcing system, where cheerfulness begets more cheerfulness.

Here's an example: Researchers instruct a group of people that they must choose from a selection of "effortful" tasks. These tasks include fun, happy-making jobs and more neutral undertakings. Which would you choose? The most socially meaningful task? The most planet-saving task? The healthiest task? Well, if you're a cheerful person, you'll choose the task that makes you happy. This probably seems like a no-brainer to the Extraverts in the crowd. *Duh! Why would you do anything else?* But we would, we Neurotic and Conscientious and Agreeable types. We have a hundred ways of not enjoying ourselves that never even crossed your mind.

Another example: Researchers assemble a series of photos showing various facial expressions: sadness, fear, happiness, and "neutral." They present them slide-show fashion to a group of people, asking them to identify the expressions. "Normal" people identify the expressions predictably. But cheerful people see the faces through rose-tinted glasses: They are much more likely to judge neutral expressions as the face of happiness. So even in a gray world, the cheerful personality finds reason to smile and wave. *Everyone else is happy, too!*

And denial helps—a powerful, biological form of denying memories. Psychologists call it "repressive defensiveness." It means you're defending your own psyche by repressing negative memories and emotions. Those dark old thoughts simply don't come up. Unimpeded by gloomy recollection, you proceed on your merry way. Cheerful people do this automatically. They don't even know it.

There's no guarantee that a cheerful Extravert will not slip and fall down the spectrum to a less beatified position. I've seen it happen. A card-carrying Extravert, cheerful and optimistic, was battered by a few hard knocks in a row. Her optimism is gone. I don't know if it's temporary or permanent. It's going on three years now, with no sign of reversion to the original arrangement. It's proof again that genes aren't everything. The most gladsome genes in the repertoire can fail to buoy your personality if your environment is the equivalent of cement shoes.

Evolution of Cheerfulness

In some ways, Extraversion shares a single spectrum with Neuroticism. It's the approach end of the approach-avoid dichotomy. In fact, research into human happiness often reaches this quirky conclusion: It's not so much that you're happy. It's that you're not Neurotic. Yep, the factor that most strongly predicts your cheerfulness isn't high self-esteem, a million friends, or a house on the beach. It's a low level of Neuroticism—anxiety, depressiveness, self-consciousness, and so on.

And who wouldn't be glad to be free of that burden? The Neurotic personality is all about avoidance, fear, scary faces, bumps in the night, tigers under the bed. The cheerful personality merely...isn't! And it's great!

Of course, it's useful, as well. Especially when times are tough, it's useful to charge forward without fear. After all, tough times are often a local problem. An elephant died and poisoned the water hole. A fire destroyed the clan's forest home. General Motors is laying off a thousand workers. The Neurotics aren't good for much under these conditions. We sit in the house, glued to the Bad News channel, seeing no good options.

But the cheerful Extravert does not take local conditions personally. The cheerful Extravert straightens his necktie and marches out to find a clean water hole. He talks to strangers as he goes. In his brain, their wary expressions are rendered welcoming, as if by magic. *Yeah, elephant died in the water hole, how about that? There's a river that way? Thanks, buddy!* Another satisfying social interaction, another squirt of pleasure from the dopamine system. The Extravert's brother was killed a few years ago when he was exploring this very region, but those dark memories rarely rise to the surface. The possibility of failure doesn't cross his mind. His pain tolerance is high. He's looking forward.

Neurotic personalities operate the opposite way. Try as we might, our prefrontal cortex is unable to quash the wailing of the amygdala. It will be heard.

Today the difference in how the two personality types handle hard times is as relevant as ever. Avoidance and approach are about as fundamental as animal behavior gets. Each carries risks, each carries rewards. A starfish who detects a rotting fish wafting on the currents must decide whether to approach the opportunity but possibly be attacked; or to avoid exposing itself but possibly starve. Approach. Avoid. Both styles work in the long haul, but you can't be both at once. It's best to have a few of each in the family.

Consider those General Motors layoffs. The people who will find new work first are likely to be the positive, energetic Extraverts. Sure, they hit a lot of dead ends, hear a lot of the word "sorry." But they take it well. It doesn't discourage them. They're not weighed down by the memory of a sister who spent four years looking for work, or the rising cost of corn futures, or the Great Depression. They get up in the morning, straighten their necktie, and march forth again.

Extraverts tend to put more effort into job hunting, and find work faster. And there's even some evidence that talking to strangers helps a person find work faster. *Hey, buddy, do you know anyone at RightCorp?* Some evidence suggests that such social networking is more effective than gathering masses and masses of data on your own.

Oh, great! I'd like to know what issues are being kicked around the water cooler before I interview. I told you about my previous water cooler, right? Yeah, dead elephant.

Extraversion Facet:

ASSERTIVENESS

ASSERTIVENESS INDICATORS	RARELY	SOMETIMES	OFTEN
I ASK FOR WHAT I WANT	❏	❏	❏
I TELL PEOPLE IF THEY'RE BOTHERING ME	❏	❏	❏
I STICK TO MY GUNS	❏	❏	❏

THIS GIVES YOU A QUICK look at where you land on this facet. If your answers tend toward the "often" side, you're higher in that facet.

People with high assertiveness stand up for themselves. When their rights are in jeopardy or their opinions are under siege they push back. They stand their ground. They believe they have the right to express themselves. And they defend their interests and disagree with others without surging into aggression.

For example: Aggressive is the student who lurks inside the school door, tripping kids as they cross the threshold. Assertive is the student who picks himself up, tells the staff what's going on, and goes to class.

People low in assertiveness aren't so sure they have a right to be heard. And they lack the confidence to put their view forward as though it's as good as anyone else's.

Like many of the Extraversion facets, assertiveness is about moving forward, approaching, even if it's only to the extent of defending one's boundaries. And also like many of the Extraversion facets, the inverse of high assertiveness looks a lot like Neuroticism, the urge to minimize one's exposure and reduce one's risk.

Assertive Mouse

We recognize animal assertiveness more quickly in more complex animals, such as dogs. I am the not-so-proud owner of a bully dog. Left to his own devices, Kuchen will stalk strange dogs at the park, closing the final yards with a mock charge. Out of respect for other owners who don't know he's bluffing, I generally call him to heel until he's too close to work up the drama. But when he does get a chance to intimidate dogs, their responses are quite diverse. Quite human.

Some dogs, usually younger males, will whip around to confront Kuchen with an attack of their own: *Groooooowl!* He has invaded their space, and now they're going to invade his. That's a form of aggression. That's the kid who comes in the school door, is tripped by the bully, and jumps up to punch him.

Some of Kuchen's targets, often puppies, cringe or flop onto their back. Even if dogs could speak, words could not be more eloquent than this. This is the international gesture of surrender. This is the kid whose only goal is to gather his scattered belongings and slink to the safety of his homeroom.

The remainder are assertive. Some of these confident dogs turn to meet his aggressive posture with a casual wag. Others actually approach their erstwhile aggressor, wagging confidently. And still others let him progress right through his charge, ignoring him. They hold their ground, but without anxiety and without aggression. And poor Kuchen, his bluff called, undergoes an awkward transformation. He might raise his head to the "friendly" position and wag like an idiot, or veer off as though suddenly captivated by a blade of grass.

That spectrum of responses is common in nature. Animals often use posture, or body language, to resolve their conflicts. By communicating at a distance, they can often avoid physical conflict, and the loss of fur, feather, and blood that ensues. Those who telegraph a willingness to defend themselves, through assertive posture, are less likely to be harassed than those who look more reactive.

Mice deploy a similar lexicon, familiar to people who watch them closely. When an aggressor comes forward, a mouse may rear onto his hind legs and turn up his nose. This is defeat. But a more assertive mouse will turn sideways to present a larger silhouette: *Take heed, brother. I'm not running.*

These postures are a far cry from the nuances of human assertiveness. But they do instruct us that the behavior is related to defending yourself when you're provoked. And in that sense it does mesh with other dopamine-driven behaviors. Moving forward to protect yourself and your territory is essential behavior for any animal. How readily you make that move—always, or only in dire circumstances—is what differentiates Extravert mice from introverts.

And from Neurotics. Neuroticism, recall, is about retreating in the face of danger. The opposite of an assertive animal is the one who turns tail and flees when a bolder mouse swaggers into his territory, or noses around his mate, or eats his lunch. It's the Neurotic mouse. It's the Mini-Me.

Assertive Human

Human research on assertiveness is limited to pure psychology, usually psychology of the workplace. It's not what you'd call "hard science." Nobody is sliding assertive people into MRI machines to see what the brain does when they say, "I'm sorry, I was sitting in that chair," or, "Excuse me, the waiting line starts back there."

Perhaps the biology underlying assertiveness is a drive to protect the things one owns. For most animals, this would be a territory, along with the food, water, shelter, and family members therein. Perhaps what we call territoriality in other animals we call assertiveness in ourselves.

I don't mean territorial aggression. Aggression carries the threat of physical violence. That's the behavior that a resident mouse uncorks when an intruder crosses his boundary. I'm thinking of territory maintenance. When wolves mark trees and rocks with scent, they're not frothing with rage. They're just restating a fact: *Ours.* And there is no rage when I trim my front hedge. I'm just asserting my boundary: *Don't walk here, don't sit here, don't park here.*

And people range tremendously in their assertiveness. I am one of the least confrontational people I know. I am outwimped only by my friend "Celexa," who described her anxiety disorder in the previous chapter. Experts on danger, we err on the side of short-term safety. When we're confronted with a challenge to our rights or territory, our amygdala wail like a flock of fire trucks and we dive for cover, hearts pounding. Rather than tangle with you, we'll let you eat our lunch. We hope that the whole encounter is so unremarkable that you forget we exist.

When I asked my assertive friend Robin how asserting herself differs from being aggressive, this is what she replied:

"It's about getting what you want, standing up for yourself. I think to be assertive you have to be confident. I think it's got a lot to do with confidence being paired with extraversion.

"You don't need to be confident to be aggressive. In fact, you might be aggressive precisely because you don't have confidence. There's nothing threatening about being assertive, but there's plenty of that about being aggressive. I think the latter is about intimidating someone, getting something from someone.

"Also, it's got nothing to do with being angry—although I'm not afraid to confront people who are bugging me. All I can say for sure is that it's fundamentally got nothing to do with aggression or anger. I actually don't have much of a temper."

Evolution of Assertiveness

If we think of assertiveness as your willingness to defend what's yours, as territoriality, then it appears to be a crucial feature of personality.

Let's go back to Mitzi and Maxi Mouse, as they would be in the wild of my backyard. Mitzi, as we saw in the chapter on Neuroticism, is an avoidance-oriented mouse. She's risk averse. She shies away from new situations, and retreats at the first sign of danger.

So how does she fare when an intruder female comes sniffing around on a fine summer's eve? This is an unfamiliar female, she knows that by smell alone. And Mitzi has a nest full of babies to consider. Never mind the seeds and insects she's accustomed to harvesting in the few square yards around her nest. If the stranger is just passing through, if she just takes a few seeds, Mitzi and her brood will survive.

But what if the intruder detects Mitzi's babies? Usually a wild mouse female will kill another female's pups when she discovers them. It's a quick and easy way to reduce the competition her own pups will face. What's a Mitzi to do?

If she asserts herself with the approaching stranger, she risks everything. If she loses the confrontation, she could lose not only her pups, but also her own health if the stranger seriously injures her.

On the other hand, if she avoids the confrontation, slips away from the nest and hopes for the best, then she loses only half the battle. Her pups will die. But Mitzi will quickly breed again. That's not so catastrophic. Avoidance is a reasonable response.

Unless the stranger intends to stay.

If that's the case, you'd want to be Maxi. Maxi is out foraging when she smells the intruding female. Automatically, she rattles her tail in the dry grass. *I guess you didn't see me. That's my chair.*

The stranger veers away. She's just a young female, after all, moving across the landscape from her mother's territory. She knows she's trespassing, and she has no desire to fight. She just wants to find a peaceful plot of ground where she can have pups of her own. Maxi's assertion risked confrontation, but preserved all of Maxi's seeds and offspring. Approach is a reasonable response.

Each reaction to a threat will work brilliantly—under some conditions. Under other conditions each will fail. Neither works so well that evolution makes it the universal personality. Neither works so badly that it vanishes from the repertoire of personality, buried under a pile of carcasses.

So You Think You Might Be an Extravert

Well, congratulations on making it this far. It's a physically dangerous personality type to own. As I write this my beloved spouse is gimping around the house in a neck brace. On a whim he steered his dirt bike toward a double jump and gave it gas. When he landed on the side of the jump the bike reeled up the safety berm and then over it. My spouse and the bike parted ways there, with the spouse arcing headfirst between a pair of trees and into the ground.

"I'm sorry," he groaned from the couch a few days into his recovery. "I shouldn't have done it."

"Well, you have an impulsive personality," I sighed.

"I know. That's why I have to organize things so that I'm not in a position to make that kind of mistake."

But it's a big world. It will be hard to novelty-seeker-proof it. What's more, the average Extravert has a hard time learning from his mistakes. He learns really well from his successes. To change an Extravert's behavior, it's more effective to praise the things you want to encourage. You'll see much quicker results if you reward an Extravert with cookies than if you squeak and bite him. And that's part of why he's so gung-ho. His brain prefers to remember the good stuff. An Extravert's hindsight is rose-tinted.

Unfortunately, this leaves him open to repeating mistakes. Many weekends my spouse heads out dirt biking with his son. Most times he returns bruised and bloody and spends the next few days hobbling around, rubbing some sore part. "I rode too hard. I'm just going to watch him next time," he vows. But if the pain is gone by the next weekend, so is the memory. And so is my spouse.

Emotional resilience is one of the traits I admire most in him. The bad luck, ill will, and foul manners that sometimes come his way would reduce me to a quaking lump. He shrugs off criticism, scolding, and vitriol, bouncing right back to his outgoing ways.

A more moderate novelty seeker will benefit from her inclination to explore and experience without risking so much of the physical damage. Such a personality is curious, inquisitive, and interesting, and maybe even safe to share a checkbook with.

Then there are the low-impulsive personalities like mine, which are . . . well, they're just dull. Reliable, rational, careful, and dull. But cheerfulness, at least, is one of the easier personality facets to change.

How great would that be? You could really make your personality happier. On a permanent basis.

I have to confess I had started doing this on my own, not knowing what I was up to. I started going to a 12-step program to deal with some old behaviors I was ready to get rid of. The messages that float around these programs started to penetrate:

> I can control only my own attitude, only my own behavior.
>
> I can choose to not fret over things I can't control.

Son of a gun, it started to work. The old amygdala still startles and begins to yell: *He looks angry! You're not working fast enough! The world is ending!*

But increasingly, my prefrontal cortex flexes a new muscle: *Pipe down back there. We're trying to have a nice time.*

It takes a lot of work to override a worrying brain. But it can be done. I'm doing it. What I learn now is that I'm developing something called "mindfulness." And mindfulness is scientifically proven to make you happier.

It seems to come down to living in the moment, as the old saw goes. Science says it's those moments that we notice, pay attention to, which determine if we see our lives as happy or not. And the present moment is usually much nicer than the junk our amygdalas want to ruminate over. If we can spend more time in the present and less time in the amygdala, we stand a better chance of feeling like cheerful people.

Where to find this marvelous mindfulness? It's all around you, under a variety of names. The oldest is "transcendental meditation." The Buddhists originated the concept of meditating in order to bring a calming discipline to the mind. "Mindfulness-based stress reduction" is one of the newer monikers. This strain grew out of medicine, as a method for treating patients with serious depression or pain. It has conquered the world. A quick Google of "mindfulness" will deliver a quadrillion hits, including the Center for Mindfulness, at which site you can search for teachers and programs in your area.

So you, too, can have increased empathy, self-compassion, and forgiveness. Perhaps more relevant to the Neurotics, you can have less rumination, lower stress, and actually rank lower on tests of whether you have an anxious personality. That's major. That makes me happy.

There may be yet another road to happiness, too, a method that involves no effortful exercise or self-talk. It's temporary, but effective. You may be able to clamp a chopstick between your teeth and be on the road to a sunnier and more contented you! True! A German team had human subjects hold a chopstick horizontally between the teeth (which produces a smile of sorts); or tuck it vertically under the upper lip (walrus expression); or hold no chopstick at all. Then, with the people's heads wired to measure electrical activity in the brain, researchers had them play a computer game. Errors were made. Normally, that would cause an electrical dustup in the prefrontal cortex, as the brain kicked itself for its stupidity. And in the brains of the "walrus" and "no chopstick" people, that was the case. They made mistakes, their prefrontal cortices scolded them, and they continued. They continued more carefully.

Not the chopstick smilers! The rewarding effect of their dopamine system raised their spirits even when they screwed up. Their PFCs made some attempt to belittle them and sober them up, but the chopstick was mightier than the PFC. The unwittingly happy chopstick smilers barged ahead with the task, making tons of mistakes. Their brains were too cheerful to care much about accuracy.

This builds on older research probing the classic question of face-making. It's firmly established now that smiling and laughing both brighten your outlook, temporarily. Even if you're smiling or laughing at nothing. (Howling, on the other hand, has been proven ineffectual. At least for humans.)

I think it's only a matter of time before some entrepreneur fields a device you wear in your cheeks overnight. You waken each day so cheerful and Extraverted that you float out of bed and crack your head on the ceiling.

3
FACTOR:
AGREEABLENESS

AGREEABLENESS FACETS
Trust
Cooperation
Sympathy
Altruism
Morality

THE CAST

Mascots . . . . .Introducing the
Rat and the Prairie Vole

Neurotransmitters . . , ...Oxytocin and Vasopressin

Brain Region . . . . .Back to the Amygdala

AGREEABLENESS

"THAT'S OK!"

"Oh, I'm fine!"

"Whatever you want is fine with me."

If you would rather swallow a little discomfort so that everybody else stays happy and relaxed, you're Agreeable. A person with high Agreeableness can relegate her own self-interest to the back of the bus, if that keeps the social wheels turning smoothly. She is also trusting, generous, and attuned to the people around her. But she may have trouble drawing boundaries and protecting herself from exploitation.

"That doesn't work for me. If that's what you want, I'll do something else."

A person with low Agreeableness prefers his own observations to conventional wisdom. An independent thinker, he's unlikely to get caught up in popular movements. He's more concerned with the facts of the crime than the motives; more with justice than explanations. This personality can seem hardheaded and aloof, even selfish. Low Agreeableness is not the same as "disagreeable." Rather, it's a low interest in conformity for conformity's sake, and a high tolerance for others' discomfort.

This factor relates to the human need to balance our social lifestyle with our individual needs. Everyone has to attend to both in order to thrive. But each personality finds that balance in a slightly different spot along the spectrum. Some of us sacrifice self-interest in order to maintain a strong social network; others are willing to fly without such an extensive net, if it means they can get their own business done.

Two of my favorite females, one a high-school teacher and the other a writing instructor, tied each other for "Least Agreeable." They proceeded to get in an argument about what this meant about them. Or they thought they got in an argument. The exchange took place via email, and both of these women are so forthright that it's easy to mistake their directness for anger. In fact, these two people rarely get angry, and neither holds a grudge.

Thinking about their personalities now, I have to admit that their disinterest in couching, framing, softening, qualifying, and otherwise diluting their opinions is one of the things that makes them so easy to be around. And fun to be around. The world would dress a whole lot better if we all took a low-Agreeableness person shopping with us. When you ask them how a coat looks on you, they tell you.

Shopping with a high-Agreeableness person like me is not advised. Loath to offend but incapable of deceit, I respond to pointed questions with paralytic gasping.

THIS GIVES YOU A QUICK look at where you land on this facet. If your answers tend toward the "often" side, you're higher in that facet.

Trusting . . . Rat!

Corals have a tidy means of mating. They do not rise up from their reefs to mingle in the warm seas. No, they stay put. And on an appointed night all the individuals of a species simultaneously blow their eggs and sperm out into the water. Those cells mingle in the moonlight, while the adults resume their tidy lives.

Ah, if it were so simple for other animals! Most animals must touch in order to reproduce. And touching runs contrary to the average animal's self-interest. The average animal is solitary. The average animal defends a territory by himself. The average animal wants nothing to do with his fellow animal.

There are many reasons to avoid touching. Animals are violent, even to their own kind. They might also carry contagious diseases or parasites. Females are especially vulnerable, since they're usually smaller. They may be forced to mate by a male who's genetically undesirable. There are good reasons to avoid even getting close. A stranger might kill your offspring, or steal some groceries as he passes through your territory.

As a result, most animals live solitary lives, steering clear of others. Think about it. Does the skunk patrol your lawn in the company of friends, or solo? The grizzly, the polar, and the black—there are no buddies among the bears. Cougar, jaguar, bobcat, tiger, margay—all together now: Solitary! Salamanders, snakes of all sizes, snapping turtles, spotted newts, and horny toads, none care for companionship. And that is the norm. Animals are mean. Even to their own species.

Hence the necessity of trust. Some force must overpower that instinctual fear if a solitary animal is to reproduce. And so some animals have evolved an entire dimension of personality that allows them to touch. We call it trust or Agreeableness, but ultimately it's "antifear." It's chemical courage that permits an animal to accomplish the deed of mating. Not to mention the deed of raising loud and demanding offspring. We take it for granted that animals enjoy the company of their offspring, and maybe even their mate. But they don't, deep down. The brain modifications that allow mothers to bond with their infants, and fathers to bond with the mothers, were major advances for life on earth. And in humans, who dwell for long periods with offspring, mates, even unrelated neighbors, those modifications have been further modified, to produce behaviors that stop us from hitting and kicking our family members.

"Reproduction and parenting: I would say complex brain structures evolved to fulfill these complex social tasks," asserts Inga Neumann. "I believe the development of this intense reproductive behavior is also a starting point for very complex behavior in many mammals."

Neumann runs a laboratory a few cities east of Klaus-Peter Lesch's, in Regensburg, Germany. She's in the Agreeable business, studying the way brain chemicals change a rat's mating and parenting behavior. I wouldn't say that a ton of Agreeableness has rubbed off on her, though. She stopped returning my emails after conceding to a date, but not a time, for my visit. I never heard from her again, and found her lab only by wandering the vast campus, asking everyone I met. As I approached her door, it swung open and she started out. Saw me. Her face fell. She backed into the room and sat down behind her desk.

Even fallen, it was a beautiful face—fine boned, with big blue eyes under feathery blond hair. She wore sleek blue jeans and a belt with a rhinestone buckle. Pretty chic for a scientist. She was otherwise all business. I'm not saying she was low in oxytocin, but it's possible.

Oxytocin has long been considered a female hormone. In rats and humans alike it floods through a pregnant female as she nears childbirth. It causes the uterus to contract, ejecting offspring. (Doctors administer pitocin, a synthetic version, to induce delivery.) Oxytocin floods anew whenever the offspring nurse. It causes milk to flow. But it works above the neck, too.

"Released into the brain and the blood, it causes birth and milk release in the female," Neumann says. "And at the same time it stimulates maternal behavior, to make sure the female *wants* to nurse her offspring."

And lately Neumann's team has been looking at what oxytocin does to the male brain. This is tricky—in either sex. If you want to know how a chemical alters behavior, you need two things simultaneously. You need to vary the level of that chemical in the brain; and you need your animal to continue behaving so you can monitor the effect. That explains the beanies worn by many of Neumann's rats.

When she shows me to the main lab, we walk into the middle of a brain surgery. A large man is hunched over a white rat who, anesthetized, is being fitted with brain dipsticks. The surgeon slices daintily through layers of pink skin and yellow fat over the skull. At his elbow the rat's newborn pups slumber in a rosy pile on a heating pad. Neumann picks up one to display its stomach. Almost a quarter of the animal comprises a white bubble: milk.

With a tiny drill, Neumann explains, the surgeon will make a pair of holes in the mother's skull. One will hold a tube through which oxytocin can be delivered to a precise location in the brain. The other will hold a teeny dialysis probe so small it can harmlessly extract fluid samples from another precise location. Finally, screw holes will allow him to stabilize a protective beanie over the skull with jeweler's screws. Stoppers will plug the top of the implanted tubes. Then, during experiments, tiny hoses will connect those tubes to machinery, for real-time monitoring of brain chemistry. After surgery, the rats take no notice of their beanies, or the tubes that sometimes rise from their heads to an instrument outside their condo.

This work was not always so orderly. Neumann tells me she started her career in East Germany before the wall came down, when no one could afford equipment. She built her own microdialysis probes. Let me repeat that: *She built her own microdialysis probes.*

"Oh, it was horrible, horrible," she shudders. "Either the science wasn't working or my one-year-old son was sick. I was always leaving the institute crying. But I didn't give up. I was fascinated by the brain."

On another bench another patient is recovering from anesthesia. Rats bounce back from surgery, and they're not prone to infections. They're tougher than humans. Just a couple of days later Neumann is able to monitor the rats' real-time brain chemistry as they mother and mate and fight and perform their whole repertoire of social behavior.

Why rats? In many ways they're easier to work with than mice. The most obvious way is their size. One of Neumann's students is performing similar experiments on mice, and she's struggling with the surgery. Everything's so much smaller. But rats also learn faster, and smell nicer. (Much nicer. No contest.) They cannot, however, be genetically altered as easily as mice. To produce anxious rats you can't just knock out a gene. You must breed them carefully over time.

Various people have done this, and so you could buy an anxious rat off the rack, so to speak. Neumann, characteristically, breeds her own. Then, using her tiny plumbing fixtures, she can map the parts of the brain that release or absorb a given chemical. And different patterns emerge depending on the social challenge Neumann presents to a rat—introducing an intruder rat, or scattering the pups around the cage, for instance.

"Our expertise is, where are these neurotransmitter chemicals released?" Neumann says. "Which one is released, and where? And when? And I'm totally convinced we would have the same results in a human."

Two striking results of oxytocin research in rodents stand out for me.

One is that a mouse whose oxytocin system is disrupted doesn't trust her mate. Even if she mated with him recently, she treats him like a stranger who might be coming to kill her babies. So oxytocin is central to breeding, for mammals.

Second is that the amygdala, the ancient fear center, is very rich in oxytocin receptors. In other words, that particular part of the brain is extrasensitive to the calming influence of oxytocin. The chemical really does act as an "antifear" serum. Love potion? Hardly. Trust nostrum? No. It's more like a sedative that helps an animal overcome its terror of strangers.

Moreover, Neumann believes the primary function of oxytocin is to protect the brain from the hormonal havoc caused by birth and parenting. "Sexual steroids fluctuate up and down in the brain during birth," she says. The roller coaster can cause stress, anxiety, and depression. But oxytocin appears to shut down the brain's normal stress reaction, in both rats and humans. The world can burst into flame and fall to pieces around a mother with a newborn infant, and she will smile and coo and sigh with happiness. Or fearlessness.

Trusting Human

Not even Inga Neumann can build a dialysis probe that humans could safely wear in our brains as we perform our social behaviors. And you can't even measure the level of chemicals in blood to get a sense of what's going on in the brain, thanks to the brain-blood barrier. Blood vessels in the brain are wrapped in filtering cells that prevent large molecules from seeping into the brain. This isolates the fragile brain from many toxins that might be circulating in the blood. Unfortunately, it also means that blood drawn from your arm reveals almost nothing about your brain chemistry. Hence it's nearly impossible to know what the chemicals are doing inside the skull when one human interacts with another. But in 2002, scientists realized they could bypass the brain-blood barrier with some substances by shooting chemicals up the nose. Crossing the barrier, we moved a step closer to real-time observations.

It was a few years more before researchers puffed oxytocin into human noses. Then it quickly became clear that oxytocin isn't just for females. It has plenty to do with male behavior, too. Both sexes reacted by overflowing with the milk of human kindness.

At least, human kindness was the effect the experimenters meant to capture. Using the classic "Prisoner's Dilemma," they gave a group of "prisoners" twelve "money units." Each player had an option of giving some units to an unseen partner, whereupon the amount would be quadrupled. The unseen partner then had a chance to return some of the loot to the prisoner.

If you were the prisoner, how much would you risk on a stranger?

Among players who didn't snort oxytocin in this experiment, just 21 percent trusted their partner with their whole wad. Among those who did snort oxytocin, however, 45 percent threw all fear to the wind. Egad!

Newspapers and blogs spasmed with speculation: Would marketing wiz-
ards fill the air in Macy's with oxytocin then sell us titanium golf clubs or
a mink bathrobe?

Lost in the fuss over the study was the fact that all of us already have
oxytocin onboard. And that 21 percent of us had proven to be naturally
trusting enough to bet the whole ranch, even without supplemental oxytocin. Trust, underwritten by oxytocin, is a natural facet of human personality. Some of us have a lot of it; some have less.

In fact, it's those who have very low trust that are more interesting to
science. People with autism have unusually low interest in other people's
faces, expressions, and even company. In this way their brains are more
like those of solitary animals. Research on dosing autistic people with
oxytocin is just beginning, but initial results show promise: Autistic people are able to identify emotional expressions more easily after they've
huffed oxy.

Neumann thinks oxytocin may help others as well. From what she's
seen in rats and mice, she's curious to know if extra oxytocin could help
women through postpartum depression, a condition that disrupts mother-
infant bonding. It certainly appeals to common sense. If a burst of oxytocin at birth helps most women open their hearts to a loud and
unattractive new infant, perhaps an insufficiency of oxytocin allows fear
and anxiety to dominate the emotions. Neumann thinks oxytocin might
treat social phobia and other anxiety disorders for the same reason: It
calms the fevered amygdala, particularly the socially fretful amygdala.

Is this a good time to hearken back to the subject of the brain's complexity? In the Neuroticism chapter we dealt with anxiety. There, we saw
that serotonin strongly influences how anxious you are, or aren't. But we
also saw that serotonin drugs don't work for everyone. Now, with oxytocin, we see how anxiety can be caused by more than one chemical. It
stands to reason that serotonin drugs won't ease your anxiety if your anxiety is caused by the oxytocin system.

So, if you had access to an MRI machine, wouldn't you want to see
whether the average amygdala behaves differently when you administer
oxytocin? It's not quite the same as sampling fluid directly with a dip-
stick, but it would tell you if sniffing oxytocin makes an amygdala more
or less alarmed at the sight of scary photographs.

And it does make an amygdala less alarmed. Normally, seeing someone else's fearful expression should wake up your amygdala, because

whatever is frightening that person could also be a threat to you. So normally, looking at another person's scared expression causes your own amygdala to shift into a state of high alert. Its job is to watch out for your safety, and any sign of danger should push it into emergency-management mode. But in the MRI machine scientists can watch oxytocin smother the flames in the amygdala. *Calm down. Chill out. Trust.*

Evolution of Trust

How did trust evolve? Let's just say that if it hadn't evolved a very long time ago, life on earth would look quite different today. Some animals would reproduce asexually, I suppose, splitting in two like bacteria. Ouch. Others might rely on the traditional strategy of coral: Male and female toss their reproductive cells into a public arena and leave the rest to chance. I suppose some creatures might even have found a way to produce pollen as plants do, letting the wind or bees act as matchmaker.

Oxytocin is an ancient neurotransmitter chemical. Animal lineages much older than the mammals—amphibians, reptiles, worms, even the egg-dumping fish—all sport some version of it. For hundreds of millions of years animals have been practicing sexual reproduction. And for that long they've also needed a chemical pry-bar to push them into proximity.

Why do some people (and some mice) have more built-in trust than others? Why would evolution maintain a spectrum, instead of boiling us down to one model?

I would have to speculate that this variety reflects the fact that mating opportunities are not constant. All animals are driven to reproduce. The DNA must go on. But the urge to mate doesn't guarantee you will have opportunities.

Let's take two of Neumann's rats out of the lab and release them in the hills of Regensburg. One, Trusting Tess, is a high-oxytocin, trusting rat. The other, Doubting Dee, is lower in both oxytocin and trust. We'll run the girls through two years to view the benefits of each personality.

The first year is a tough year for rats. Food is scarce, plague has killed lots of rats, and there simply aren't many guys left to breed with. But one day a scruffy and skinny male appears on the scene. Tess and Dee can tell just by the scent of him that he's riddled with fleas that would love to jump from him to them. And he's so hungry he might just as soon attack them as mate with them. Doubting Dee surveys the situation and her amygdala quivers with fear: *Danger! Aggressive male! Parasites!* She lies low, letting him pass.

Trusting Tess has a different reaction. Her amygdala tries to warn her of the dangers, but her oxytocin is powerful enough to quell the alarms. She approaches. She mates.

At the end of session one, Tess is ahead. Yes, she has fleas, and her offspring are undernourished from the scanty food supply. But she has managed to reproduce, even in tough times. It pays to drop your guard and talk to strangers.

The next year the environment smiles on rats. Food is abundant and the weather is mild. Male rats are a dime a dozen. Tess doesn't do anything differently this year. She approaches the first guy on the scene, and mates. But Dee takes her time surveying the field of candidates. She sniffs each from a distance, noting his health status, his social status, and even whether he's too closely related to her. Only when she has singled out the healthiest, most aggressive rat does she offer to share her uterus.

This year, Trusting Tess will again have small pups, because once again she has caught fleas from an unhealthy male, leaving her anemic. But Dee, still skeptical, has waved off many potential suitors. She hung back until she had identified the biggest, baddest male available. Her pups will be large and robust. Her sons will presumably inherit their father's willingness to compete for mating privileges. It pays to be suspicious of strangers.

Humans have more complicated social lives than do rats. Trust plays a part in many more of our social interactions. But the fundamental dynamic is probably the same as it is with Trusting Tess and Doubting Dee. Some of us are quick to grasp at social opportunities. Others are more wary, lending an ear to our amygdala: *Be careful, be afraid.*

Those of us who are trusting, who don't take time to look under the hood and kick the tires, may pay for that "good deal" later on when we're stranded on a lonesome highway. And those of us who are inclined to look a gift horse in the mouth may hem and haw until the donor decides to give his horse to somebody a little more grateful. Each tendency has strengths and weaknesses. The fact that nature has kept the whole range in our gene pool argues that each tendency works just fine, over the long haul.

COOPERATION

COOPERATION INDICATORS	RARELY	SOMETIMES	OFTEN
I GO ALONG TO GET ALONG	❏	❏	❏
PERSUASION IS BETTER THAN FORCE	❏	❏	❏
I LIKE WORKING IN A GROUP	❏	❏	❏

THIS GIVES YOU A QUICK look at where you land on this facet. If your answers tend toward the "often" side, you're higher in that facet.

Cooperativeness measures a person's willingness to compromise. This version of cooperation is more about setting aside your own agenda than whether you work well with other people. A personality with high cooperation comes across as flexible and easygoing. Low cooperation describes a more independent person who isn't afraid to confront a group when she doesn't share its goals. Rather than falling into step with the majority, she's inclined to speak up and argue for her own course. A low-cooperative personality can work perfectly well with a group, if the other members don't roll over and give up.

Cooperative Rat

So you got close enough to another animal to mate. Congratulations! If you're a male, you can go now. Your work here is done. If you're the female, you will combine eggs and sperm, slowly wrap each pair in a membrane, and add fat and nutrients as you seal them up. Sometime later you'll lay the eggs and then you, too, will split.

That's the old method, anyway. Lots of animals still practice this form of reproduction. Most snakes carefully conceal their eggs and slither on with their lives. Frogs and toads generally lay and leave. A female iguana might hang around her nest for a few days to be sure no one digs it up, but that concludes her maternal obligations.

Somewhere along the line, however, a few animals evolved some extra maternal aggression. They would stick by the nest through the entire incubation phase. As the eggs hatched these new-style mothers would continue to guard the young, ripping into any predator who approached.

The first time I saw an alligator pottering about in a Florida swamp with a baby alligator on her head I gaped in dismay. I assumed the young one had mistaken the adult's head for a rock, and crawled up there to sunbathe. Any moment I expected the big gator to toss her head back and gulp down the little one. Slowly I realized this was not your average reptile. How odd! For a predatory reptile to recognize a bite-size animal as her offspring seemed so...unnatural. After all, a variety of snakes, fish, and lizards will snack on their own young with no apparent misgivings. What was going on with this alligator that her instinct to capture food was constrained?

Cooperation, that's what. If trust allows two animals to get close enough to mate, it's cooperation (that's the human term anyway) that allows an animal to compromise her short-term interest so that she and another animal will both benefit. I suppose an alligator might call it tempering: She tempers her prey drive in favor of her genetic legacy.

It's not conscious, this tempering. It's one brain system overpowering another. And it came about entirely by genetic accident. But the accident worked. An ancient animal harbored a genetic mutation that caused her to stand by the nest as her young crawled out. Her extended vigilance saved more of them from predators. Her daughters who shared the mutation also kept more of their own offspring alive. Thus behavior spreads through the gene pool.

What genetic mutation was it? In rats and mice, at least, it apparently involved a chemical called vasopressin.

Around a corner or two from Neumann's office I find her colleague Oliver Bosch. A brick of a guy with buzzed hair and wire-rim glasses, he tells me how one measures mothering in a rat. He's a rare researcher whose fondness for his animals is unconcealed.

"What we call 'blanket posture' is a position where the mother is spread out to cover the pups, just like a blanket, to keep them warm. When a dam is arched-back nursing, she's standing over the pups and she's really arching up her back so they all can nurse. 'Licking and grooming' is time the mother spends taking care of the pups. And to test pup retrieval we scatter the pups in the cage and measure how long it takes her to bring them back to the nest." These are the tests of motherhood, if you're a rat. And they require human monitoring.

"You spend hours and hours and *hours* in front of the cage," Bosch sighs. "But it touches me. I become a father every day. A father to tons of pups."

What Bosch learned from watching mothers in action is that vasopressin makes moms anxious, and that anxious moms are good moms. He and Neumann began by temporarily disabling the vasopressin system in the brain of rat mothers. They injected a blocking agent through an implanted tube, then they monitored the mothers. Before their eyes, good mothers lost interest. They didn't arch their backs as often for their pups to nurse. And they spent less time with the pups overall. "They provide the minimum of maternal care," Bosch says, "but nothing extra."

In a second round of testing, the vasopressin-deprived mothers also put less effort into retrieving scattered pups. "They almost didn't retrieve the pups at all."

Not content with this evidence, Bosch and Neumann flipped the experiment: They put *more* vasopressin into some rats. Under the skin of some mothers they implanted a tiny pump. The size of a drug capsule, each pump fed the brain a steady supplement of vasopressin for a week. In went vasopressin, and out came exemplary mothering. With extra vasopressin these mothers arched their backs like so many Brooklyn Bridges, standing motionless for the convenience of their pups.

"Arched-back nursing seems really important," Bosch says. "The less the mother does, the more anxious the pups are as adults. And their social skills are reduced." So good nursing technique, inspired by both oxytocin and vasopressin, helps to shape a young rat's personality.

The recurring question, of course, is: If it's true for rats, is it true for us? Does human mothering behavior rely as heavily on vasopressin as rat mothering does?

"It would be really nice if this could be used in humans," Bosch says. "Postpartum depression—maybe vasopressin is a way of treating this. But we don't even know if the vasopressin system is active in humans—which isn't so easy to study." He sighs, studying his desk for a minute.

"It's so important to have animal models. It's possible to really measure neurotransmitter release while the animal is showing the behavior. It's possible to take out the brain right after the behavior so you can see what was going on at the moment. But it's hard." He looks up. His face is tight.

"The mothers are easier because you need—you need that brain tissue. But it's really hard to kill pups." He pauses. "They can't live without the mother. I know that night I will have a nightmare. And that's important. It keeps you on a level where whenever you have to do it, you think about it."

Vasopressin isn't just for mothers. Just as oxytocin was long thought to be the "female" chemical, vasopressin's prior reputation was for producing the opposite of cooperation in males. One of Neumann's young PhDs, Alexa Veenema, has shown how easily a disrupted childhood can alter both a male's vasopressin system and his aggression. Simply separating male pups from their mother for three hours a day was sufficient to make them a whole lot less cooperative as young males. By the time they were old enough to "play fight" with other males, these guys were already bullies. They bit their peers more, took the submissive role less, and were generally on the attack.

So we can't say vasopressin is a cooperation chemical. But if vasopressin doesn't make a male more cooperative, what does? Well, vasopressin! But only in animals that must behave cooperatively to breed successfully.

For a demonstration, we'll jump from rats to yet another rodent, voles. Voles are like mice with shorter tails. Like mice, they come in dozens of species: There are water voles, snow voles, sagebrush and heather voles, even a Duke of Bedford's vole. Most species mate like rats and mice, with whoever is around at the time. The sexes don't share parenting duties, although some males may hang around the neighborhood to keep an eye on their belongings, as it were.

And then there is the prairie vole. Gray-brown on top and tawny below, tiny of ear and wee of eye, the prairie vole lives in the dry lands of the Canadian and American Midwest. It inhabits a system of tunnels and trails among the prairie plants. And it inhabits this domicile in couples—prairie vole couples who cuddle and groom each other, who stick together not just through one round of offspring, but right on through their lives.

Lest this get too sentimental, a shot of reality: Prairie voles "mate for life," but they are not purely monogamous. Like swans, humans, and other animals that mate for life, your average prairie vole also is willing to diversify its offspring when an attractive opportunity presents itself. This is cooperation we're talking about, after all, not altruism.

So what does a male prairie vole do with his vasopressin? Has he evolved to be aggressive on this chemical? Or does it make him go all maternal and mushy, as it does a female rat? How does he exploit this reproductive agent?

Well, first of all, he has relocated it. With the flip of a gene, just a tiny alteration, the male prairie vole has shifted the location of his vasopressin system. Merely by lowering his vasopressin center a few millimeters in his brain, he set the chemical to a completely different task. Instead of helping him fight off males, it helps him cooperate with a female.

The vasopressin effect kicks in after he mates with a female, or even if he spends a couple of days near her. Henceforth he prefers to be in her vicinity. The two voles huddle together and groom each other. But his vasopressin daze does more than help him tolerate another animal's presence. The male actually helps with domestic chores. The two animals work together to build a nest and guard it from strangers. The male will huddle with the pups to warm them, and if a baby toddles out of the nest he'll retrieve it just like a female.

Experiments to block vasopressin in the male vole's brain produce a stark contrast: He'll mate, but he won't cooperate. Without vasopressin in just the right place, he's chemically unequipped to remain with his family. Conversely, adding vasopressin to the prairie vole's brain makes him even more attentive to his mate. The montane vole, by comparison, cannot be forced into domestic partnership through manipulation of his vasopressin. His vasopressin receptors—locks—aren't in a part of the brain that regulates cooperative behavior. You can pour a pail of the stuff into his head and the montane vole will stay his solitary course.

But perhaps one could add those receptors where one wanted them? That was the question researchers had in mind when they made an attempt to transplant prairie vole cooperativeness into mice. They copied a stretch of vole DNA that includes the vasopressin receptor gene, and patched it into mouse DNA. These test-tube babies did in fact receive a personality transplant. Male mice matured into very attentive fellows. When introduced to females, they spent twice as much time as normal mice on sniffing and grooming them. And in their brains, the vasopressin receptors had shifted downward, to prairie vole territory.

The question hovers in the air, so urgent and fluttery I can sense it: *Could you do this to a human male?* I mean, I can even sense pronouns: *Could you do this to my husband?*

Cooperative Human

The prairie vole and the human are among the tiny minority of mammals who bond when they mate. The rarity of this behavior is enough to prove that it's a tough row to hoe. If it were easier to co-parent than to go one's own way, more animals would do it. It's an unusual behavior necessitated by special conditions. If you're a bird migrating four thousand miles to raise kids in the brief northern summer, you want to make sure your babies grow up in time to carry the family name back south in the fall. If you're a human, your babies are so slow to mature that a single parent is hard pressed to provide food and shelter until it's independent.

Female mammals already have some cooperativeness built in, thanks to oxytocin. Females are forced to compromise their short-term goals in order to care for their offspring. If they weren't built to do that, they'd saunter away in search of their own next meal, and perhaps never come back. Like snakes.

But male mammals lack mammary glands and the chemical infrastructure that goes with them. The human male has been forced to improvise his own method for staying in the room with another human for years on end.

Apparently they, like male voles, have come to rely on vasopressin. That's the suggestion of a recent study, anyway. In a humbling parallel to the prairie vole, the quality of a human husband appears to be influenced by which variation of the standard vasopressin gene he has onboard.

The gene in question is the same one the prairie vole altered to become a snuggler: It regulates the placement and number of the vasopressin receptors. Like many genes, it comes in a few different models, of varying lengths. And one model, now the infamous 334, doesn't seem to work very well. According to a large Swedish study, men who received copies of 334 from both their parents were particularly nonbondy. Of the married and cohabiting men with two copies of 334, one-third had experienced a marital crisis or near-divorce in the past year—double the normal rate. They were twice as likely to be cohabiting, as opposed to married. Even males carrying just one copy of 334 were more likely to report that their marriages weren't cuddly. And their wives were likely to concur.

But the human male has multiple uses for vasopressin. In their "mating" relationships, it seems to allow them to rein in selfish behavior in order to work with a partner. But their relationships with other males are another matter, according to researchers who puffed vasopressin up the noses of their subjects.

Before administering vasopressin the scientists applied electrodes to the little muscles that pull the eyebrows together in a frown. These muscles react to your emotions without your knowing it, sometimes in ways too small even to see. But the electrodes can detect the electric current that triggers those muscles. Thus wired, the men were presented with photos of strangers. Male strangers.

With extra vasopressin up their noses and in their brains, the men scowled at angry faces, scowled more at happy faces, and scowled furiously at neutral expressions! When asked to consciously rate each face for friendliness, men on vasopressin saw nothing but potential challengers, and they micro-scowled for all they were worth. *I'll take you all on!*

Another study found that men hopped up on vasopressin got more stressed out during experiments designed to make them feel awkward in front of others. It looks as though the human male has evolved to use vasopressin to determine how aggressively to treat strange men.

The team also puffed vasopressin into women, giving us a glimpse of what it does to the human female. The team showed these women photos of women's faces. As you might expect of this reproductive chemical, it had quite a different effect on the female brain. On the women's faces, a different muscle, the zygomaticus, twitched in response to all the portraits they were shown, whether the portrait was angry, happy, or neutral. The zygomaticus, when fully contracted, pulls the lips away from the teeth, producing an expression humans call a smile. On other primates the expression is a "fear grimace," but the message is the same: *Don't attack me. I'll cooperate.* Contrary to men, when vasopressin-puffed women consciously analyzed the expressions, their brains detected more friendliness than was really there. So in same-sex relationships vasopressin makes males more suspicious and aggressive but makes females more trusting and friendly.

Oddly, a handful of studies argue that vasopressin afflicts the sexes equally when it comes to … song and dance. A little background: Theorists have long nursed a hunch that music and dance both evolved as aids to human reproduction—but probably not in the way you think. Both these rhythmic behaviors helped mothers to calm offspring so that the wailing wouldn't attract predators, and the mother could get some work done. *That* part of reproduction. From there song and dance broadened into mating displays that advertised a person's physical and intellectual quality. That's the theory. And the theory is now supported by a trickle of research showing that certain versions of the vasopressin receptor gene turn up more often in professional dancers and people with musical talent.

That same gene also relates to the Prisoner's Dilemma, which tests how much money a person will invest in an invisible partner. In both men and women, the trusting investors and the skeptics tend to cluster around particular variants of the gene.

The fact that vasopressin seems to influence how social a man will be has made it a popular focus of research on autism. Autism is an overwhelmingly male condition, and it's marked by a disinterest in social bonding of any sort—even with parents, in some cases. Autism is probably one of those disorders that, like schizophrenia or depression, involves a different assortment of genes and chemicals in each person. But both vasopressin and oxytocin appear to play some role.

Evolution of Cooperation

I can more easily see how cooperation would evolve in females than males. The female mammal, whether rat, vole, or human, is compelled to stay near her infants so they can nurse. So, that's one layer of cooperation she needs. That bondage means the female must neglect her food foraging when she's nursing and must leave her young unprotected when she's foraging. But if she's able to tolerate a female sister or neighbor, the two can share child care and reduce their risks. That's a second layer of cooperation.

Compared with most mammals, the human female is especially burdened by offspring that require constant attention for many years. For the human female, to gather food and keep the home fires burning while raising a toddler alone would be an impossible task. For that matter, the human female has a hard time even giving birth unassisted. So presumably, the human female has long been wired to make and keep allies.

Male mammals don't experience the same pressures to bond. They don't need to tolerate offspring clambering over them. And forget about tolerating other males. Every male rat is in competition with every other male rat, to mate with every possible female. That's hard soil for cooperation to sprout in. Thus when mammals do live in groups one aggressive male usually reigns as dictator. If there is cooperation between males, it's often the chimpanzee kind: *Hey, Buddy, the Alpha has been harassing us both. Let's team up and get rid of him.*

Somehow our own species overcame that natural animosity that divides males. It may be that males evolved to bond closely to a mate, like the prairie vole males. Thus bonded to one, they weren't in competition for all the rest. Who knows how we did it. But we did: Human males are uncommonly tolerant of one another and of group living in general. Females still tend to rank higher on this personality trait than males, but is it any wonder? The female human is part of a biological tradition of tolerating offspring that stretches back more than one hundred million years. The human male . . . isn't.

For both sexes of our species, the benefits of cooperation must have outweighed the costs. Presumably we evolved living as small groups of primates who shared food, child care, and sentry duty. It seems obvious to me that a group that included a majority of cooperative individuals would thrive. And a group heavy with rogues and mavericks would fly apart at the first hardship. The scattered members would wander the wilderness and be eaten by wolves. The cooperative humans, who shared hunting, parenting, and fighting duties, would multiply and take over the world.

If males lag somewhat today, it's not by a lot. In my sample of friends, a male ranked highest on cooperativeness, and a female ranked lowest.

Obviously the species retains some diversity of cooperativeness, spread on both sides of the gene pool. As always, that implies that each setting on the cooperativeness dial has a function. Let's see how each personality type, from doormat to stone wall, is useful.

For this test let's again send a couple of rats out of the lab and into the wild. Female rats, like female humans, are cooperative breeders. Multiple females will sometimes share a nest, and share some parenting duties.

So meet Cooperative Cate, who has opted to share parental chores with a bunch of other girl rats. The most obvious benefit is that Cate can forage for food longer than if she were a single mother. Someone will feed her pups and retrieve them if they wander off. Other mothers will fight off snakes and strangers who might want to kill her pups. But she also benefits from the many noses and ears watching over the collective young. Like the crows who will forsake their territory in my backyard to join the neighbors in pursuit of a hawk, animals are more formidable when they stand (or fly) together.

But cooperation is not risk free. The same food Cate seeks is sought by all the other mothers too, so it goes fast. She has to work harder to keep her pups in milk. Furthermore, one of the females has a lot more pups— the cooperation is out of balance. Yet another mother has a cold virus, and she's giving it to all the pups. Some will die. And when the snake does come sliding around, are the other mothers going to defend Cate's pups as vigorously as they defend their own? I doubt it. Worst of all, some mothers in the group will spontaneously kill other females' pups.

This is why Maverick Mary chooses a solo path. On her own, Mary has to leave her offspring alone when she forages. Nothing stands between them and the foxes, skunks, raccoons, crows, and even other rats who might discover the nest. But Mary's pups are safe from the murderous and diseased mothers in the cooperative group. And food is more abundant with fewer rats around, so her foraging trips are shorter. And every ounce of her effort is devoted to her own brood, not someone else's. If she's a strong and healthy rat, she may be able to raise an exceptionally healthy family.

And what about the fathers? A male low on cooperativeness won't be motivated to hang around in either scenario. His reproductive drive will spur him onward to mate with additional females. And if some of his offspring die from lack of food and protection, it's a small loss for him: His strategy prioritizes quantity over quality.

If he's a cooperative rat who can tolerate the proximity of a partner, his strategy will switch to quality over quantity. He'll only father one litter, but these pups will benefit from the attention of two full-time parents.

Risks and rewards, rewards and risks. Each of these strategies, for each of the sexes, has its strengths. The fact that rats and people alike display a range of this personality facet is proof.

I expect cooperative behavior is especially suited to small communities where you are related to many of the members. When you're related to the other mothers, their offspring will carry some of your own genes. The energy you spend caring for their infants actually promotes your own DNA as well.

That math holds true in both rats and people. But while many rats do still live in small groups where cooperation pays this genetic bonus, that is a dying lifestyle for humans. With genetic distance comes an emotional distance, too. The less we know about people, the less we're willing to cooperate with them. Being anonymous lowers your risk: If none of your own comrades are around, you can turn your back on a hungry child without damaging your reputation for cooperativeness.

THIS GIVES YOU A QUICK look at where you land on this facet. If your answers tend toward the "often" side, you're higher in that facet.

The questions measure the bleeding-heart aspect of sympathy, the aspect that makes a person compassionate, humane, merciful, and a total softy. A person with high sympathy feels other people's emotions deeply, and is moved by them. A low level of sympathy, in contrast, conveys some immunity to the suffering of the world. A low-sympathy personality tends to consider a dilemma from a cool distance, as opposed to rushing in on a wave of emotion.

Sympathetic Mouse

Mice and rats aren't supposed to have sympathy for each other—let alone empathy. Tradition has it that you can't empathize or sympathize with another unless you first have a concept of your own self. And most researchers haven't been able to catch the twinkle of self-awareness in the eye of a mouse. Chimpanzees, yes. Dolphins and elephants, yes. But mice?

Actually, even such an ethereal quality as sympathy appears to be rooted down in the earthy genes. And as you'd expect of a personality trait, some mice are more empathic than others. There are nice mice, and mice who are more aloof.

Overseeing a lab full of such mice is Garet Lahvis. I mentioned that I encountered some unusually sympathetic researchers while researching the Agreeableness factor. First Oliver Bosch nearly wept over the sacrifice of his rat babies. A couple of weeks later Lahvis, a behavioral neuroscientist, was sighing over the emotions of his mice.

I find him in a hilltop lab in Portland, Oregon. He's tall, dark, and floppy, in a sleepless and unshaven way. He's been up for three days finishing an important grant proposal. It's due in minutes, and he's emailing it back and forth to a colleague between frantic trips to a printer in a distant room. A Tums bottle stands on his desk, empty.

"We write these grants to keep the research going," he murmurs, staring at the computer screen. "You start the work ahead of time, but you don't know what you're going to say. So you put it off—" The phone rings. He grunts into it, then strides out into the hall again, sweatshirt billowing.

He thought I was coming tomorrow. But he'll neither send me away nor let me take him to lunch. "No, this is good. I'll decompress."

He spent the first fifteen years of his career studying environmental pollutants, he says. "But I got really excited about watching animals do things together. I was convinced that animals have a rich emotional life. Every time someone says, 'An animal can't do that…'" He grins. He swings his chair back to the computer and plays a video file: Two mice meet in a plastic condo and begin the usual sniffing survey of each other. But this isn't an old silent film. This is a talkie. Lahvis has recorded the mice and modulated their ultrasonic squeaks down to human hearing range. As the mice greet and explore each other, they issue bursts of birdlike chirping. It's an endearing little stream of chatter. Lahvis smiles.

"I don't *know* what that's about, but I've got an idea!" he says. He hurls himself backward in the chair until he's nearly prone, hands locked behind his head. "I think it's a pure, clean expression of an emotional state. When you feel bad, you're in a low-energy state. When you're feeling good, there's a bounce in your step, and your voice goes up and down. You chirp more when proximity makes you feel good."

But proximity doesn't make everyone—or everymouse—feel equally good. After Lahvis and one of his researchers, Jules Panksepp, began eavesdropping on mice, they tested two strains to see if they were equally communicative.

They used young mice, which are naturally more social and playful than adults. And they tested the two popular research strains against each other: Black 6 vs. BalbC. It was the Inaugural Mice Nice-Off.

I must interject that ounce for ounce it is hard to beat the cuteness of a mouse. For such a tiny animal, there is a huge amount of cuteness packed in there. But perhaps an even better bargain is young mice. Their snouts are stubby like puppy snouts, and their eyes are big. When two meet, each uses his stubby paws to grip his comrade, and they poke their noses into every place they can be poked. At this age even males can be housed together and will not fight.

Lahvis & Co. took advantage of this and separated the roommates, each of whom would spend a day in isolation. At test time, experimenters reunited the roommates. They measured the time the mice interacted as they greeted each other, and they measured the amount of chirping each produced.

If you're looking for empathy, tell your sad story to a Black 6 mouse. By a long shot, they showed more interest in their roommates. And they communicated their interest with a babbling brook of mouse chirps. The BalbC mice weren't disinterested. They, too, gripped and sniffed their roommates. But in some cases Black 6 mice spent nearly twice as long on their greetings. The only difference between the two strains of mice is their genes. Which genes? That's a mystery for now. Whichever they are, they're making a large difference in how sociable the mice are.

Hmmm, thought the world of mouse emotion. *Perhaps there's something going on here...*

Lahvis & Co. pushed the animals a little further. It's one thing to be social. A school of fish can do that. Empathy is a taller order. Technically, empathy is "feeling into," it's catching someone else's emotional state like you might catch a cold.

(Sympathy, technically, is *acting* on someone else's emotional state. Your crying makes me sad—that's empathy. Your sadness makes me want to hug you—that's sympathy.)

This time the researchers put a BalbC and a Black 6 in a gallery adjacent to a foot-shock chamber. Strange mice came into the foot-shock chamber. Scientists played a thirty-second warning tone followed by a two-second foot shock. The gallery mice could not see into the shock chamber, but they could hear the shocked strangers squeak and hop around. As usual, the shocked mice quickly learned to associate that warning tone with a shock. When they heard it, they would freeze.

Now the gallery mice had their turn in the shock chamber. And before they were shocked even once, some of them heard the warning tone and froze. Never having experienced the shock, they had caught fear from the shocked mice. They had absorbed the others' fear, and it had altered their own emotions.

Some of them had, anyway. Not all of them had gotten the message. You can probably predict which strain was most sensitive to fear in other mice, based on the experiment on how the different strains greet each other. The same mice who squeak and chirp and nose each other—the Black 6 mice—are the ones who learn from their comrades' scary experiences. Apparently, they're more socially sensitive.

"We think this is a more sophisticated case of empathy," Lahvis explains. "We think this is *feeling into* the feelings of another individual."

He sighs. Spins the chair a little. Lahvis is the first scientist who tells me up front that he tests the foot shock on himself before he subjects mice to it. It's the mildest jolt that all mice will respond to. If it were milder not every mouse would react to it, and the data would be fouled.

"This research can be difficult when I see that these animals have emotions that resemble emotions I have myself," he says. "It raises a lot of hard questions."

But he stays in the mouse-shocking business for the light it sheds on autism, schizophrenia, and other disorders of the personality. An autistic mouse, for instance, would neither pay attention to, nor learn from, another mouse's pain. By studying the differences between empathic and not-so-empathic mice, you might find drugs or gene therapies that could awaken empathy in an autistic child. But he's also learning about the emotions of creatures we've always assumed were emotionless.

"People who focus on humans say humans are the only social species," he says. "I think the real value of this work is: How do you treat animals? It's the environment. Imagine all these salmon going up the Columbia, who are used to a tumultuous river. They may feel like crap now because we need dams and electricity, so the salmon get these stagnant pools."

He offers to drive me back down the mountain to town. Panksepp, a tousled young man, is coming into the lab as we go out. On a shelf over his desk stands a book called *Sleep: A Comprehensive Handbook*. It sure looks comprehensive. It's four inches thick.

And you wouldn't guess from the looks of these two guys that either of them has read much of it yet.

Sympathetic Human

A friend got divorced recently, and also got taken to the cleaner by her lazy ex. One night a gang of us were discussing the whole mess, and this friend, "Patsy," was enumerating reasons her ex should stop hassling her for more money. She had brought most of the furniture into the union, and had now left much of it behind. "I gave him the huge new TV," she said. "And the entertainment center. And my bedroom set. A really nice bedroom set. And the couch."

Robin, whose general Agreeableness score was quite low, began to steam as the list grew.

"I felt so sorry for him," Patsy continued. "I didn't want to leave him with an empty house."

Robin boiled over. "Sorry for him? Where is he? I'm going to get that furniture right now."

"He doesn't have a job ..."

Robin snorted. Then she slyly inquired, "Hey, if *I'm* mean to you, will you give me furniture, too?"

Probably. Patsy recently went to Sears and purchased a long winter coat for an immigrant woman she had seen collecting bottles and cans in her neighborhood. Patsy doesn't tolerate the suffering of others very well. It makes her suffer.

How? The mice haven't revealed the biological secret to their newly discovered empathy, let alone why some have a lot of it. But humans have produced a few clues as to why some of us are bleeding hearts and others have hearts of stone.

Not surprisingly, oxytocin appears to be a sponsor of sympathy. Returning to the (putative) origins of all Agreeableness, a Dutch team has looked at how oxytocin influences parenting styles. They watched as a bunch of mothers interacted with their two-year-old children, who were trying to solve a difficult puzzle. Some mothers were patient and helpful; others were not. And the not-so-helpful mothers were more likely to carry a particular version of the oxytocin receptor gene: Their "mommy chemical" system may have been set just a tad to the selfish side, slightly blinding them to the emotions of their children.

Now further studies are finding that oxytocin can increase the amount of money people will donate to a charity. One study in particular lent credence to the time-honored method charities use to pull money from magazine readers: Feature a woebegone child in your advertisement. In the study, researchers had subjects watch a tearjerker film of a father talking about his son's brain tumor. They sampled subjects' blood before and after the film. Following the film the blood was awash in oxytocin, and their donations to charity rose 47 percent, compared to those of subjects who saw a film of the same father talking about a trip to the zoo. The tearjerker technique was more effective on women than men.

Experiments wherein people sniff oxytocin to bolster the chemical in their brain show that the chemical may work in two ways. It may operate first by dampening our natural fear of one another. Oxytocin is very active in the amygdala, which monitors the world for danger. Extra oxytocin fights fear.

Then, with that terror out of the way, perhaps it's easier to read another person's emotions and relate to them. People dosed with oxytocin make more direct eye contact, and they are better at describing the emotions portrayed on another's face. So extra oxytocin also helps us to empathize.

But humans have access to another brain system that raises sympathy, too. When you stick out your tongue at a baby, the baby will often stick its tongue out automatically. The motor region of the baby's brain is mirroring your own motor region. Our emotional regions also have a system that helps us to mirror another's feelings. Although many scientists refer to this system as "mirror neurons," referring to brain cells that reproduce other people's emotions in our own brain, that's speculation.

Mirror neurons do exist in monkeys, that's established. When scientists monitored one nerve cell at a time to see how one monkey responded to a second monkey's actions, they found that some neurons fire just as if the watching monkey were performing the action himself.

Whether a monkey reaches for food or merely watches another monkey reach for food, his neurons fire identically. Scientists can confirm mirror neurons in monkeys because they're allowed to slip superfine wires into a monkey's brain and tap in to one cell at a time. They can't get a permit to do that to humans.

Patsy, the giver of furniture, and I sometimes joke that we have too many mirror neurons. For bleeding hearts like us it's a struggle to keep other people's emotions out of our heads. Your pain is my pain. Everyone's pain is my pain. I've learned to watch only happy movies, and to avert my eyes from advertisements for animal charities. If I didn't throw a blanket over the mirror neurons I'd spend the whole day in tears.

When researchers use MRI to hunt for a mirror neuron system in humans, they argue over what they see. Compared to monitoring a single cell with a wire, MRI yields a grainy picture. There is plenty of room for interpretation. Some think a couple of structures in the middle layers of the brain behave in a mirrorlike way. But others see two distinct types of neurons—one for watching, another for acting.

That we have some kind of mirroring system is common sense. A means of automatically mimicking another animal's behavior would speed the learning process. And certainly it would help to explain how one animal can feel empathy for another. It would also explain how effortlessly, subconsciously, we empathize with a sad face or a happy face. Your mirror system would reproduce inside you the emotions you saw in someone else.

Right now we can say that oxytocin seems to make a human more sensitive to others' emotions, but we can't say how.

Evolution of Sympathy

That Dutch study of how mothers helped their children is central to the evolution of sympathy. On its face, it seems so cold and sad that some mothers can't take their child's point of view. It seems so unmaternal that they scold instead of praising, that they dominate instead of guiding. Human mothers, we're often told, are selfless creatures dedicated entirely to the health and welfare of their offspring. They're not snakes who give birth and glide away. These selfish women must have no empathy, sympathy, or decency!

But that ignores the fact that a mother—every mother, whether snake, skunk, or sheep—has biological aspirations above and beyond an infant. In her DNA she dreams of launching not one, but a dozen offspring down the river of time. And to do that she has to watch out for her own health and welfare.

All mothers and their infants engage in a battle over this issue, from the moment of conception. It is in the offspring's best interest to drag every nutrient and calorie it can absorb out of its mother's body. It is in the mother's best interest to hold something back so that she can raise future offspring. This battle continues after birth. An infant denied the opportunity to nurse does not quit without a fight. She'll let loose wails that in earlier times could attract deadly predators.

But no matter how sympathetic a mother might be, the infant won't gain the upper hand in this contest. Starvation remains a real threat to humans today, and the photographs that come out of refugee camps testify to the importance of motherly selfishness. Still strong enough to walk, mothers embrace their dying children. I'm sure they wish they could nurse their children, but evolution has outfitted them with bodies that will not permit it. When a female's body fat drops below a certain point, she can continue to empathize with her child's pain, but her body refuses to sympathize: Her body stops producing milk.

And how could it be otherwise? Why would evolution reward a body that would give its last calorie to an offspring, then die and leave the offspring to starve alone? The offspring of such sympathetic mothers don't survive, and neither do the genes that would make a person so disastrously generous.

But neither could evolution produce mothers who would abandon an infant at the first sign of hardship. Evolution rewards those mothers who invest in their existing offspring but guard their ability to have more children in the future.

And the dynamic would not be much different for men. Human infants are so useless that they require twenty-four-hour protection for a number of years after birth. Many hands make light work, and they also make for more surviving babies. Males who are inspired to pitch in with child care increase the odds that their own genetic legacy will grow healthy and strong.

The contrast between Patsy and Robin highlights the fact that humans come in many shades of cooperativeness. Why would that be? Why does evolution perpetuate both the pushovers and the pushers?

Well, a personality that's low in sympathy or empathy is not a heart-less block of stone. She just isn't so quick to assume the feelings of others. She does a better job of maintaining her boundaries and keeping a steady eye on her own future. Undistracted by life's melodramas, she's more likely to focus on facts and figures.

Nor is the bleeding heart (ahem) a boundaryless ball of mush. Well, maybe she is. Maybe it's a real challenge for her to say no, because she can feel the impact of that hard word on another's psyche. Maybe she's sucked into drama after drama because she cannot walk away from a soul in distress. But because she is what we think of as "a good friend" she also has a large circle of humans that ensure she has a healthy balance of laughter and martinis in her life.

THIS GIVES YOU A QUICK look at where you land on this facet. If your answers tend toward the "often" side, you're higher in that facet.

Altruism is cooperation on steroids: It's tolerating others to such an extent that you actually sacrifice for their benefit. A personality with high altruism enjoys being helpful. He holds the door for you and helps the old lady cross the street because it feels good. A personality with low altruism doesn't get that warm glow. He holds the door because his mother trained him to, and when he sees the old lady standing helplessly on the busy corner, he walks to the next block to cross. Perhaps he has calculated that it's more efficient to spend his time earning money and writing a check to Friends of Old Ladies.

Altruistic . . . Voles!

Sticking around to see a child through the first few years of life is plenty challenging for an animal that's fundamentally selfish. Mating for life—that calls for ropes and chains. And since the female mammal is already bound to her offspring by oxytocin, the target here is the male. Most male mammals limit their parenting to contribution of the male reproductive cells. It takes more than a home-cooked meal to convince your average male mammal to call a single location "home," and a single female "sweetheart."

It takes vasopressin. And a lot of it. Enough to overcome the bias present in every animal: that the "correct" behavior is that which brings him or her immediate and positive results. Rambling off to mate with a dozen other females, now that's productive behavior. If you're struck by lightning at the end of the day, you can go to the happy hunting ground knowing that you gave your best effort.

On the other hand, staying near the first female you mate with, helping to collect grass for a nest, watching over vole babies—that defies short-term logic. If you're struck by lightning halfway through that process, your solitary mate may be unable to compensate for your lost labor. Your babies may be eaten by a snake. Game over.

We encountered vasopressin in the discussion of cooperation. It makes a female rat anxious and supermaternal, but it makes a male prairie vole mate for life. And as we saw, scientists can transplant the prairie vole's self-sacrificing personality into a mouse by giving the mouse the vole's vasopressin gene.

But is mating for life a form of altruism? Altruism isn't easy to recognize in animals. To qualify, a vole must perform an action that benefits another, but not himself. In humans this behavior includes such acts as giving money to a charity, holding the door for a stranger, or even rubbing your spouse's back without the expectation that the favor will be returned.

Sometimes what we think is altruism in animals is an instinct gone haywire. For instance, a dog might adopt a kitten or a lion cub. But the dog is a domesticated animal whose instincts are eroded by breeding, and orphaned animals are primed to bond with anything that happens by.

That said, scientists have documented altruism in a handful of species. A vampire bat that spends a fruitful night bloodsucking will share with a less successful cave mate, regurgitating blood for her hungry comrade. (Gacking up blood many not sound altruistic to you, but vampires appreciate it, I assure you.) Various dolphin species will support a sick dolphin on their backs so it can breathe at the ocean's surface. And sometimes in lab experiments chimpanzees and other primates will spend some effort in retrieving food for a friend even if they don't get a treat themselves.

But voles? To say that a male vole is altruistic when it helps to raise its own offspring is a bit of a stretch. By that definition all female mammals are massively altruistic.

The male vole does go a step further, to bond with a female. The female does likewise. This step is a big one. It makes sense to bond with and support your blood relatives, because their success carries a percentage of your own DNA forward. But a mate carries none of your DNA (you hope). Even the smallest favor you perform for a mate, if your favor doesn't also benefit your kids, is altruism. It's giving more than you get. It's biologically odd, and special. Do male voles do that?

Perhaps we could say this: A carefully placed vasopressin gene makes a male prairie vole behave less selfishly than your typical vole. Rather than sowing his wild oats and spreeing around the neighborhood, the prairie vole invests all (well, most) of his energy in hearth and home. In addition to protecting his offspring, he grooms and huddles with his mate. Does that qualify as altruism? If not, it is at least unusual. Vasopressin, as expressed in the male prairie vole, seems to create a male mammal who can put immediate self-interest on the back burner in order to raise his offspring and give his partner the equivalent of a back-rub.

If this discussion seems to neglect the altruism of the female prairie vole, well, it does. Hearkening back to cooperation, remember that female mammals have enjoyed the benefits of oxytocin for many million years. That chemical allows a female to tolerate and even bond with offspring. For her to extend her ancient bonding ability from pups to a mate is not such a leap. Predictably, the female prairie vole does rely on her oxytocin, not vasopressin, to facilitate her marriage.

Altruistic Human

Oddly, fifteen years passed between the discovery that vasopressin wrought monogamy in male prairie voles and the news that it has a similar effect in the male human. I would have moved that question to the top of my pile the day after the prairie vole news was announced.

Instead, scientists who sought the roots of human altruism have gone looking in both sexes. And often, they've explored through the lens of that old money game, the Prisoner's Dilemma.

The Common Goods game is similar, but designed for groups. It reveals how you might respond if a member of your group refuses to behave altruistically. It starts with each player holding some money—let's say five people each holding $10. The scientist informs players that they may each make an anonymous investment in the "common good" of the group. The scientist will double the resulting pool of money. Then she will divide the total amount equally among the five players, regardless of who gave what.

What would you do? I would start by putting in $10. I'm a hopeless altruist. I'm also hopelessly rational: If everyone puts in $10, we'll all get the maximum reward. Duh! Of course I'd put in $10!

Maybe you wouldn't. Maybe you and another player would put in $5. And the two remaining players would put in $0. Now the researcher doubles the $20 total to $40 and gives us each back $8. Everyone gets back more than they gave—except me.

OK, time for the second round. What will you do this time? You can tell from the payout after the first round that some players were more selfish than you were. But you still got back more than you gave. Will you give $5 again, even though your generosity will probably benefit someone else more than it will benefit you? Even when someone is sitting on his cash and letting you risk yours?

Researchers have run this experiment a zillion times, in a zillion different cultures. Culture does influence the results somewhat. For instance, in a culture where offering charity is insulting, players give little. But the general outcome mirrors the Prisoner's Dilemma. The average person is willing to commit half of what he has to the public good. A few people always freeload on the system, getting something for nothing. And a few others always throw all their money in. Humans come in varying degrees of altruism.

When experimenters tinker with the Common Goods game to illuminate sex differences, their findings aren't surprising. Women often contribute more, although they quickly adjust to the prevailing strategy of the group. They're also quick to cooperate on that prevailing strategy, so that contributions and payouts are stable even if they're not fair. Given that female mammals have a much longer history of sacrificing for their offspring than males do, I don't find this shocking in the least. Altruism comes easier to females, I imagine, because a female's brain is more robustly wired for sacrifice. In any mammal that has evolved a group lifestyle or a monogamous mating style, the male brains would also be wired for cooperation. But their wiring would be relatively young.

Of course, scientists are curious whether a person's degree of altruism correlates with his vasopressin setting. Human vasopressin genes, like many others, come in varying lengths. Shorter versions tend to make weaker vasopressin systems, and long ones make more dominant systems. Would that translate, as do short and long versions of a serotonin gene, into different personality types?

Recently a team zeroed in on the human version of the vasopressin system gene that affects male voles so powerfully. They tested the relevant gene in 203 college students to see which lengths each person carried. And then they set the students down to a test similar to the Prisoner's Dilemma. How much money would each student give, and would their generosity reflect their genes?

The graph is as tidy as you like: When averaged together, people with two copies of the short version of the vasopressin gene give the least; those with one long and one short copy are middling; fools who give the shirt off their back are most likely to have two copies of the long version. Vasopressin and altruistic behavior look as though they're bound in holy matrimony. Or in biological matrimony, at any rate.

This is a rare study that also analyzed human brain tissue to solidify the connections between genes and behavior. Although working with laboratory mice entails a fair number of rules and regulations, procuring cadaver parts—even a small chunk of brain—is a lot more arduous. But once procured, the brain samples were eloquent. Again researchers checked the length of each brain's vasopressin gene. Then they measured the actual brain tissue to determine how active each person's vasopressin system had been in life. And again they found that the length of the vasopressin gene did correlate with how active the vasopressin system had been in each brain. Interestingly, these researchers found no difference between men and women.

Of course, that old spy, the MRI machine, has granted many more glimpses inside the human brain when it's grappling with altruism. Two brain regions, one ancient and one new, come online when a brain is challenged to put its immediate self-interest on hold.

The old part is the limbic system, which helps us feel good when we behave in ways that will keep the species going. Fueled by dopamine, it activates to reward us for eating and breeding. It gives us the happy giggles when we win the lottery. But it also pats us on the back when we do good deeds. It does feel good to do good, and that's because of the limbic system. It's heartwarming that the brain gets just as big a charge from giving as getting. Some research says the brain is even more excited by altruism: It really is better to give than to receive.

The other important brain region is that rational CEO, the PFC (prefrontal cortex). When older parts of the brain issue a knee-jerk response to a social situation, the PFC steps in to be sure we don't put our foot in our mouth. After all, we're social animals, and it's important that other people like and support us.

Thus, MRI studies have found that the PFC comes online when altruism bumps up against self-interest. Here's one scenario used to test a brain's reaction to that struggle: When I give you $10 to stick in your pocket or donate to a charity, what will you do? Do you donate less when I'm watching than you would if your decision were private? How will you behave if the charity is controversial—right to die, or abortion—and my opinion is the opposite of yours? With so many factors to weigh, the PFC takes the helm.

Tellingly, some people who have suffered an injury to the PFC seem inured to a sense of guilt that motivates many altruistic acts. They can pursue their self-interest without the tempering influence of guilt and self-doubt.

And the fact that women are more altruistic than men, on average, takes me back to that Swedish study of men and the "good husband" vasopressin gene. The men who had two copies of the not-so-bondy 334 version were twice as likely to have faced a near meltdown of their partnership in the past year, compared with the other men in the study. Without biological support from their brains, they seem to have a harder time sacrificing for their marriage. Meanwhile, the researchers found that it didn't matter which version of the vasopressin gene a woman carried. Women evidently rely on a different system to produce their altruistic behavior.

If that one study is any indication, the human male is not completely domesticated. There remains a wild streak in the male genome, a steely insistence on taking the solitary route. I'm neither surprised nor disgruntled by this. Rather, I think it's grand when science affirms the general observation of the general populace. The world over, male and female mammals evolved to perform different tasks. It's self-defeating to expect identical behavior from the two sexes.

Evolution of Altruism

I would rather spend my limited money on bleaching my hair than feeding a starving African child. I prove that every three months. I mentioned my choice to my hairdresser.

"It's a matter of distance," she said. "If you had to step over a starving child to get in here, you would take care of the starving child instead."

That sums up the current thinking on how altruism evolved. It evolved when humans lived in small groups. It still works best in small groups, and falls to pieces over a distance.

From a distance, humans don't appear any more self-sacrificing than centipedes or camels. One out of six humans alive today gets less food than she needs. This is not because humans are unable to produce enough, but because we opt not to share. The cost of protecting a child from malaria with a sleeping net is $10, but three thousand children a day die of the disease. Even in the richest nations of the world, children go hungry, and adults with terminal diseases live outdoors in the wind and weather. From a distance, we look like a heartless bunch.

The truth is we don't feel very effective when we try to help from a long distance away. If someone told me that five children in Africa would die this year if they didn't get $10 mosquito nets, I'd be able to imagine my $50 protecting those five children. But tell me that my $50 can save only five out of the three thousand who are predicted to die today, and even after I donate, tomorrow's entire daily allotment of three thousand will die... ugh. It's overwhelming.

But come a little closer. Recently my state debated whether to allow two people of the same gender to bind themselves into a legally recognized pair. I can't think how this would affect my life, for good or ill. So... what do I care? If I were a solitary species, devoid of social impulses beyond those required for mating and child-rearing, I wouldn't give this issue a second thought.

However, I know a lot of people whose lives would be greatly improved if they could gain legal recognition as a family unit. I know people who yearn for that, who are pained that they do not now have that recognition. The pain of these people pains me, and stirs me to rise up from the comfort of my legally recognized household. I told one of these friends that I would help launch a fund-raising party. I wrote checks, and set about harassing my own friends to attend.

It's not a small sacrifice. We have very little free time. We were involved in a second campaign, too. We had a lawn that hadn't been mown in weeks, and a boy who would much rather go dirt biking up north than plan another stupid party. Quiet nights at home had become a distant memory. But we can't walk away from these members of our community. When push comes to shove, we scrounge up the time and the money that will help them.

And don't we feel like exemplary citizens! We've sacrificed! We've reached out a helping hand! We've gone above and beyond! It feels great! When we see friends who share the cause, we feel bonded by our common altruism. We recognize one another as fellow members of a special tribe: We are helpers. We help.

Why? Again, why? Helping does not help us in any detectible way. Even less does it help our friends who quietly wrote a check and never came to the party. Does the fox spend an extra hour hunting, exposing itself to coyotes, in order to carry a rabbit to its arthritic neighbor? Never, never, never. That would be very bad fox policy. That would contribute nothing to his own offspring's welfare, and in fact would bolster the strength of the arthritic neighbor, who competes for the same rabbits.

Charles Darwin wept over the puzzle of altruism. On a finite planet, life, too, must be finite. Survival is clearly a matter of competition. Each organism is always scrambling for the food, water, shelter, and mates it requires. The fallen seed that can deploy its root quickly and unfurl a leaf will suffocate a sibling that's slower to sprout. If instead the fallen seed restrains its own growth to permit its sibling a head start, then it will be overshadowed. The genes that caused its generous behavior will be extinguished, lost from the pool. Darwin couldn't cipher how niceness could evolve, since only a gene that allows its host animal to triumph will be passed along.

Lately scientists have studied the puzzle of altruism with games such as the Prisoner's Dilemma. And as Darwin would have predicted, it is inevitable that someone behaves selfishly and comes out ahead. By exploiting others, he thrives. Cheating pays.

But again, come a little closer. Put aside the computer games played in anonymity. Come down to our neighborhood beach in the morning, when dogs are permitted off leash for two hours.

There was a time many years ago when people paid insufficient attention to their dogs. Poop clotted in the dune grass and among the dry mounds of seaweed, adding to the human clutter of straws, candy wrappers, paper, and green hanks of lobster trap rope. Understandably, the community put its foot down. *Dogs off the beach!*

But the dog owners rallied. With a poop-scooper law already in effect, we needed to enforce it among ourselves. As we battled the city council we also formed a community of like-minded people, of helpers. Now many dog owners scoop every poop we spot, and gather human trash, too. These days no dog can poop in peace—three citizens rush at every bent backside. *A bag, a spare, and one to share!*

Cheaters still show up. They will always show up, because cheating still pays. I encountered one a couple of years ago. She had brought to the beach a fussing toddler, the toddler's stroller, and a dog. As other dog owners passed, she watched her dog climb into the dunes (forbidden) and squat. She didn't follow to pick up.

"You do know your dog isn't allowed in the dunes, and you're required to pick up?" I inquired, offering her a bag.

"I know but I can't," she said, gesturing at her child as though that expunged every transgression. She turned away. I'm agreeable, but everyone has a limit.

"If you can't take responsibility for your dog, don't come to the beach," I invited as I scooped her poop. When others asked about the confrontation I spread word of the cheater. She would find no friends here in the future. An exploiter was ejected from the game.

And that is presumably how altruism gained a foothold among groups of humans. It would have evolved when humans lived in small groups, so small that everyone knew everyone else. Each group held a few altruists, a few cheaters, and a majority of followers, who could swing either way.

Let's run the Common Goods game in a small group of early humans. We'll run it once with a preponderance of cheaters, then with a critical mass of altruists.

In the first week a few altruists throw in everything—the deer they shot, the fish they netted, the water they hauled, the firewood they collected. They stand watch for others when they're sick. The followers throw in half their goods—half a deer, a few fish. And the cheaters, who are numerous in this group, contribute little or nothing, knowing they'll be cared for anyway. *I was going to hunt but all you guys were going*... The second week a bunch of the followers keep their goods to themselves. Why bother, if the cheaters give nothing back? The bleeding-heart altruists throw in everything again. By the third week nearly everyone has joined the cheaters, which is to say that no one is sharing and there's nothing to join. The community has broken down. No one trusts anyone and no one relies on anyone. Humans can't live that way—we need too much help. Game over. The lions win.

But when the altruists face a smaller number of cheaters, the dynamic swings in their direction. With only a few freeloaders in the first round, the majority of families still benefit greatly from sharing life's burdens and blessings. This success attracts the followers to join ranks with the altruists. Giving goes up, cheating goes down, and everyone thrives. The few cheaters are recognized by everyone, and they're shunned. Here in Healthy and Happy Land, babies are well watched, and mothers well fed. The genes for generosity multiply.

I like this model because it reflects the daily reality. The cheaters I've known don't tend to attract a crowd of friends. The friends they do have may also have personalities that are hard to love. We have a lexicon for such people, and none of the terms are flattering: He's a taker. She's a player. He's a sponge, a parasite. She's an opportunist, always working an angle.

A personality from the low-altruism end of the spectrum can help a person to succeed. Cheating can in fact pay well—but only as long as the number of cheaters stays low. At the other end of the spectrum, altruists who give away the store may look suicidal, at first blush. But they're likely to have immense social support when they need it.

Most of us tend to clump in the middle. A healthy self-regard prevents us from giving away so much that we endanger our own family's survival. Although some give away more than others, the normal human will reliably help another human in need.

MORALITY INDICATORS	RARELY	SOMETIMES	OFTEN
I'M UPSET BY CHEATING	❏	❏	❏
I TELL THE TRUTH EVEN WHEN IT'S EMBARRASSING	❏	❏	❏
I DO WHAT'S RIGHT, NOT WHAT'S EASY	❏	❏	❏

THIS GIVES YOU A QUICK look at where you land on this facet. If your answers tend toward the "often" side, you're higher in that facet.

If you score low on morality, fret not. It doesn't mean you're a criminal without a conscience. As with low trust, people with low morality don't place great faith in their fellow man. They operate on the assumption that a little manipulation and some shading of the truth delivers better results than candid communication. People with high morality are more direct in their dealings. They don't beat around the bush or pad the truth. You may not care for what they have to say, but at least you know they mean it.

Moral Mouse

Mitzi Mouse would never kill Maxi Mouse. In fact, if Maxi ever had babies, Mitzi might well bustle into the nest and help her care for them. Mitzi, although she is stronger and more aggressive, does not hurt Maxi.

Quite the reverse. She grooms her, cleaning her back and head, the hard-to-reach places. An observer would be tempted to say Mitzi is kind to Maxi.

Is Mitzi a morally upstanding mouse? Is she one fuzzy ounce of ethical rectitude?

If two males replaced the girls on my desk, one might kill the other. Would that be immoral? Male mice are obligated to battle each other if they hope to mate with females. A mouse who failed to attack other males would fail to reproduce. Is that moral?

Morality is such a morass. Humans routinely kill one another in our fervor to dictate which behaviors are right and wrong, just within our own species. To arrive at a definition that also covers the behavior of fish and fowl is a recipe for global, interspecies blather.

But the whole point of this book is to examine the human personality through the simplifying lens of science. So here goes.

The behaviors we consider moral in ourselves are—coincidentally?—those that counteract our selfish urges. They are the behaviors that help us to reproduce successfully while maintaining our social organization. They stop us from killing our family and neighbors in a passing temper tantrum. They help us to harvest no more than our share from the fruit tree.

So, is it fair to say that the behavior a mouse exhibits in the name of reproducing and maintaining its social organization is also moral?

Mice, like us, refrain from attacking their mates. Call it morality, but note that it's also biologically sensible. Females, especially if they're related, will take care of each other's pups. That may seem charitable and community-oriented, but biologically it results in more protective adults hovering over each mother's vulnerable brood. These behaviors are influenced by vasopressin and oxytocin, which presumably temper the selfish urges that would otherwise prevail. And those social behaviors help mice to survive. So do mice have morals?

It is certainly possible to create a mouse who doesn't follow the social "rules" of its species. Researchers create such mice in hope of understanding the antisocial personality. Inga Neumann's lab has analyzed a strain called SAL, created by breeding many generations of the most aggressive mice. The SAL males don't posture and rear and rattle their tails when an unknown male appears in their cage. They attack. And when the researchers looked into their brains they found a deranged vasopressin system. While vasopressin helps a male bond with females, it also helps him fight off other males. And when the vasopressin is maladjusted, apparently so is the mouse.

This level of aggression doesn't necessarily help a mouse to compete. By attacking without the usual preliminaries, a hyperaggressive male forfeits the advantage of sizing up his opponent. He initiates fights that are either unwinnable or unnecessary. Some strains of hyperaggressive mice may also attack females, which is patently self-defeating: A dead female can't bear his pups, hence his violent DNA goes nowhere. Are these sociopathic mice? Immoral mice?

Humans have traditionally preferred a more cerebral version of morality. We think it has to involve a struggle with conscience. That version of morality I can't in good conscience ascribe to a mouse. Or even a male prairie vole. Mice have a lot more personality than most people imagine, but an ability to discern right and wrong is pushing things a whisker too far. Even rats, which are more social and intelligent than mice, have an equivocal capacity for self-sacrifice. A famous experiment of the early 1960s found that a rat would stop pressing a bar for food if her bar-pressing also delivered a shock to a rat in a neighboring cage. But to this day scientists debate whether that was altruism or a selfish act aimed at silencing the racket from next door.

Nonetheless, humans aren't the sole animal with an ability to judge the fairness of a situation.

Chimpanzees seem to get peeved when they see a neighbor given better treats then they. If two chimps each hand over a plastic token and one earns a cucumber slice while the other gets a grape, the cucumber recipient will go on strike: *This is wrong.*

But unlike humans, a chimpanzee makes no protest when he gets the grape and his friend gets a cuke. *I got the grape last time. Here, you take it.* No, they're fine with that kind of unfairness. This reveals a warning system in the chimp's brain that cares more about being cheated than bringing about global justice.

Domestic dogs have shown the same one-sided sense of fairness. In a famous experiment, researchers paired off pet dogs, and those pairs took turns shaking hands with an experimenter. One dog got a treat for each performance, and the other dog got nothing. The dissed dogs not only noticed the inequity, they protested it. The lead researcher's own border collie had cheerfully shaken hands thirty times in a row with no treat when she was alone. But now she looked incredulous as the dog beside her got treats and she did not. She raised her paw on demand five times; then lay down; and finally looked away, refusing to take any part in the disgraceful dealings.

It's a good bet that any species whose strategy relies on cooperative behavior will have some way of detecting cheaters. As we saw, too many cheaters can destroy a group. In order for cooperation to persist, the cooperative individuals must be able to protect themselves from exploitation.

But detecting a cheater is only half the moral equation. The other half is refusing to participate in unfair behavior even when unfairness benefits you. That's harder.

Moral Human

If you want to flex your social brain, to feel its various components strain against one another, present it with this classic dilemma. Try not to consider legal ramifications, only moral ones:

You're standing by the railroad tracks watching a runaway train come toward you. The train cannot be stopped, and after it passes you it will kill five people stuck on the tracks. But beside you is a switch that can divert the train to a spur where it will kill only one person. What will you do?

About 90 percent of people polled conclude they'd engage the switch. Aren't we good, clever, moral animals? Well, let's do it again to be sure.

This time you watch the scene unfold from the vantage point of a footbridge over the tracks, and this time there's no spur. Only a massive object could stop the train this time. At your side is an absolute giant of a man. And there's no railing . . .

About 90 percent of people would *not* push the giant. Philosophers and psychiatrists argue about what our double standard reveals about our built-in morality. But it likely has something to do with the brain circuits that screech a warning when a social animal is tempted to kill its fellow creatures. Flipping the switch has the same effect as pushing the giant. Each time, one person dies and five people live. But when you flip a switch, your brain doesn't have to envision your hands making forbidden contact with your victim.

When I insert myself in either dilemma I nearly get a headache. My prefrontal cortex considers the mathematics of the situation, the relative tragedies, and concludes that I must act. I reach out to flip the switch or push the man, and am frozen at the thought of condemning a person to death.

Morality as most of us experience it isn't rational. Research has shown that we generally declare our moral standards, then release a babbling brook of gibberish as we struggle to make our choices look logical.

Another classic brain teaser illustrates this even more embarrassingly than the train dilemma:

As adults, a brother and sister find themselves together in a city where no one knows them. They decide to have intercourse just once. They use every precaution against conceiving a child. They part ways and that's the end of it. Why is this wrong?

Again the prefrontal cortex writhes.

There is only one scientific reason that siblings shouldn't mate: Their genetic weaknesses may be compounded in their offspring. So if the two siblings are infertile, for instance, what's the problem? No harm, no foul, right? Yet the human brain squirms with revulsion at the image.

That's because the human, like the mouse, has evolved a biological aversion to relatives. It's nothing personal. Between related adults a hug and a kiss is tolerable, but nothing we'd want to make a habit of. The human brain, like the mouse brain, seems to use odor as the switch that turns off attraction. Accumulating evidence reveals that the human won't mate with anyone she or he smelled frequently in childhood. This isn't a conscious decision. It's a case of the brain automatically wrinkling its nose at any potential mate who smells too familiar. It is thus with mice. A normal mouse has no interest in mating with relatives, but if a scientist blocks her sense of smell she becomes a mouse of very loose morals.

Murder and incest are just two of the rules all humans agree on. We call these rules morals, and we argue that morality requires a conscious deliberation. But what's the difference between a mouse who avoids incest and a human who avoids it every bit as instinctively? Why do we call the human who resists the urge to kill his mate a moral man, while a mouse who obeys the same law is . . . simply a mouse?

Furthermore, why do individual humans display such varying degrees of moral conviction? I can only guess it relates to the human's ancient struggle to both compete and cooperate. Like all animals, and plants, humans must fight for their room and board on earth. Room and board are limited. Not every boy and girl who participates will get a ribbon. And so humans retain a competitive drive, a powerful instinct for self-preservation.

But our species is burdened also with ungainly offspring, and is dependent on group living for shared food and protection. Human females—and hence their offspring—are particularly vulnerable at the time of birth. Therefore that old selfish brain must be smothered, wrapped in a blanket of sociability. Humans must tend and befriend one another, or die of helplessness.

Beneath the brain's blanket of trust and altruism, the ancient and competitive brain still thrashes with the desire to kill the boy who broke your daughter's heart. Its sticky fingers reach out to fold away the extra ten-dollar-bill a clerk hands you in change. The competitive brain is alive and well in there.

The variation in human personality presumably arises because each human brain finds its own balance between those biological needs: to dominate, and to be liked. Every stratagem within the normal spectrum works. Even the most committed do-gooder will stop sharing food with the poor if her own children start to go hungry. Even the most manipulative rascal takes pains to cheat subtly enough that he won't be ejected from society.

Thus within the range of normal personality we find a person who stops feeding the poor at the point when she can no longer afford to lighten her hair. And we find others who eat only soybeans because that is what all humans must eat if all humans are to eat.

As usual when regions of the brain disagree, that old CEO, the PFC, enters the boardroom to resolve the conflict between our selfish and social needs. It's a tough job. It's particularly tough when we have to mislead— deceive, lie to—our social partners in order to meet our selfish needs. In a recent experiment, researchers slid subjects into an MRI machine and presented them with a series of coin tosses. After each toss they asked the subject if he had predicted correctly. Naturally a bunch of the subjects lied through their teeth—they reported a success rate much greater than 50 percent. But their PFCs told the truth: Their PFCs worked furiously as the players lied, striving to reconcile two opposing goals.

Scientists have shown that temporarily restraining the PFC can make a person less moral. Recently a team applied a mild electrical current to the skull over the PFC of a series of volunteers. That interrupted the region's ability to send signals. Asked now to fool an interviewer, previously tongue-tied people let fly a flock of falsehoods. A normal person's heart rate increases and her palms sweat a little when she lies. But a brain with its PFC locked in a straitjacket is able to violate moral rules with no outward signs of vexation.

This reinforces the view that the PFC is a studious juggler who manages the desires that clash in every human brain. Ever since young Phineas Gage's iron crowbar rocketed through his frontal lobe in 1848, people have suspected that "niceness" resided in that quadrant of the head. Gage's spike, launched by an explosion, returned to earth eighty feet away. But Gage may never have returned, despite a physical recovery. According to a report by his doctor written after Gage died twelve years later, the genial and accomplished young man people had known prior to the accident was gone.

"The equilibrium or balance, so to speak, between his intellectual faculties and animal propensities, seems to have been destroyed," the doctor wrote. He went on to describe a man with a toddler's brain. Gage was impatient, impulsive, stubborn, demanding, and rude. He lost his job, drifted for years, did a stint in the circus, and eventually died after suffering a series of convulsions.

I'm haunted by the report of the first doctor on the scene, who found Gage conscious and talkative. "Mr. G. got up and vomited; the effort of vomiting pressed out about half a teacupful of the brain, which fell upon the floor."

The notion of Phineas Gage's humanity lying there on the floor, that half a teacupful of brain that makes a person likeable, is chilling. It tells me that any one of us could make the same transformation, devolving in a minute from human to something less.

Evolution of Morality

To call Gage's selfish urges "animal propensities" is a slur against all cooperative species. Clearly, some animals have evolved propensities not just to cooperate with one another, but to sacrifice for one another as well.

Naturally, the scientific debate over the origin of morality is explosive. Of all the special qualities once believed to set humans apart from other animals, this would be the most unnerving one to share. Tool use, once considered a human stronghold, we now admire in monkeys who crush nuts and crows who wield rakes. Language, or at least some facility with rearranging bits of noise to produce varied meanings, now reaches our wondering ears in the chatter of starlings and prairie dogs. Morality, however, is a more serious matter. Morality we really believe to be ours.

But what if the quality we call morality is just instinct? What if the "wrongness" of incest or murder or even unkindness is merely that it's a biologically bad idea—nothing sacred, nothing more?

Both mouse and man observe a ban on killing one's mate and one's off-spring. The male prairie vole cleaves to his family presumably because to abandon them would reduce the survival rate of his offspring. The male human presumably does the same.

I dislike using different words to describe identical behaviors in humans and other animals. To call cooperative behavior "instinct" in mice but call it "morality" in men is confusing. Why not call mouse behavior moral, too? Or call human kindness an instinct? Given the baggage weighing on all those words, perhaps it's time for a fresh word altogether. We science writers try not to throw jargon around, but this is a special circumstance. Scientists have a really nice term that covers a wide range of "agreeable" behaviors: prosocial.

An animal who hopes to benefit from company and cooperation must evolve some behaviors that will counteract his fundamental instincts for competition and self-preservation. Such behaviors are known as prosocial. They allow two animals to approach each other, to work together, to share.

Different species show differing degrees of prosociality. Some, like tigers, have nearly none. A female can tolerate a male only when she needs to mate; she puts up with her own offspring only until they're grown. At the other extreme are colonial insects such as honeybees. These animals are so prosocial that most individuals never attempt to mate and pass along their own DNA. They sacrifice everything for the colony.

Humans are less prosocial than bees, but more prosocial than mice, or even prairie voles. That's clear from the large groups we form. If mice and voles were forced to live in villages of a hundred, they'd fight the whole day long. Mice and voles can tolerate immediate family and a few relatives nearby, but males especially have no need for buddies. Humans, on the other hand, do cluster in groups that include many different families.

The human male is particularly aberrant in the amount of social schmoozing he can abide. Men not only tolerate one another, but they even form strong bonds with males from different families. I can think of only a few mammals that even approximate this degree of sociality—baboons, chimps, bonobos, and some dolphins. In all these species both males and females are extremely social.

Of course, not every individual dolphin is equally prosocial. Some spend more energy looking after their own interests, and others spend a little too much time on their friendships. Some extremes don't work very well, and are weeded out. Over time and generations, the range of dolphin personality stabilizes within a range that works. And so it is with humans.

Biology has equipped us to resist the urge to kill our neighbor or steal his spear, both of which can cause a dangerous erosion of our reputation. For an animal that can't live without social support, a good reputation is so valuable that evolution has smiled on humans whose brains steer them away from antisocial behaviors.

Exceptions always turn up. The twisting and copying of DNA goes a little bit wrong, or a child inherits two copies of a low-social gene, and a personality forms that incorporates no regard for others. It happens with lots of animals. Jane Goodall documented a wild male chimp who would have passed for a serial and incestuous rapist in human society. She also knew a female who tried to steal and eat the babies of other females. Things go wrong. It happens in every species.

Humans are not tolerant of antisocial individuals. Today most cultures don't allow a person who can't restrain his "animal propensities" to drift among the rest of us. It's too dangerous. People whose selfish desires can't be contained under the blanket of their social drives often end up hurting others. These days an imbalanced person who behaves immorally can see his reproductive career fizzle behind cement blocks and iron bars.

The rest of us, whether we're a tad selfish or a bit of a bleeding heart, mind our manners. We refrain from theft and murder, we share with the needy and take care of the neighbor's children. The mouse obeys the same laws. We all practice morality for the same reason: It's good biology.

So You Think You Might Be Agreeable

Humans are a social species, like the chimpanzees, dolphins, wolves. We're too weak, and our offspring too needy, to survive as solitary animals, or even in pairs. Humans evolved in groups that presumably shared sentry duty, child care, and whatever food they could collect during the day. Animals can't rely on one another this way until they evolve methods of keeping their natural competitiveness and self-interest in check.

Individuals who were able to restrain their selfish impulses in the interest of group harmony would have formed groups that were more stable, less violent, and perhaps better fed and cared for. These tight-knit communities would have been better positioned to defend themselves and their hunting territory when another group invaded. They would have raised more children. Thoughtfulness could evolve thus.

The heart of Agreeableness is self-restraint. Whether it's a carrot or a stick that inspires the restraint is an interesting question. It may be that human pairs are also bound together by a terror of solitary life, more than a positive feeling about their union. It may be that the entire Agreeable factor of human personality is fueled as much by fear as by warm fuzzies.

Evolution doesn't care whether we respond to a carrot or a stick. Evolution simply rewards behavior that results in more offspring living long enough to reproduce. Whether an animal's blaring amygdala scares it into productive behavior, or whether its vasopressin and oxytocin seduce it into the same action, it matters not. Not one bit.

Thus we saw with Neumann's rats that oxytocin acts like an anxiety drug. Traditionally, scientists saw it as a love potion—a more positive view of this bonding chemistry. But rat brains and human brains both argue that oxytocin is a sedative for the fearful amygdala. It is by squelching fear, not fostering fondness, that oxytocin binds two animals together after they've mated.

Whatever way we come by our Agreeableness, this personality factor governs the quality of our relationships.

If you're high on Agreeableness, you probably have a lot of friends. Because your behavior is geared toward keeping other people happy, who wouldn't want to be your friend? You may even attract some miscreants who habitually cozy up to the people that give much and expect little. If high trust contributes to your Agreeableness, you may be especially vulnerable to exploitation.

Speaking from experience, a person with high Agreeableness can accumulate quite a load of hangers-on over time. It's not a bad idea to do a periodic purge. I'm thinking that next spring when I clean the basement

I'll also sort through the cape of people hanging off my neck. *Does Estelle ever call me, or is it always me calling her? Au revoir, Estelle. Tony told me twice he'd return my chainsaw and it's still not here. Ciao, Tony.*

People with high Agreeableness may also have trouble saying no, or otherwise watching out for their own needs. The yearning for social stability may eclipse personal goals. They may have weak boundaries, and feel like the doormat that everyone else steps on as they ascend to bigger and better things.

This is a personality muscle that can be strengthened, fortunately. Assertiveness training is a fitness program for the too-Agreeable brain. By exercising the inner self-preservationist, you can achieve a healthier balance between the selfish and social drives.

Even so, the bleeding heart of the Agreeable personality will keep beating. It will thud in commiseration with poor, self-absorbed Estelle whom you just fired as a friend. It will throb when you pass the animal shelter, which you will not enter because you will be unable to leave without another three cats and a blind hamster. It will clench over the beaten child, and spasm for the endangered polar bear. You can work on your boundaries, but to actually dull the reflection of your brain's mirror regions is the kind of change only a hurtling railroad spike can effect. And to be clear, that was a really lucky shot.

I can't say that life is easier for people with a low-Agreeableness personality. But it sure looks easier. My less-Agreeable friends Liza and Robin both seem to waste very little of their lives doing what I call angsting. They tend to have clear goals. They seem to put one foot in front of the other to reach them. They make good leaders because they will speak up when everyone else is hemming and hawing. Both, interestingly, are excellent teachers. I suspect it's because they don't get sidetracked by their students' dramas.

"You have to be respectful and keep a distance," Robin says. "I'm always written up as being really personable, but I don't become friends with the students." (I would be such a train wreck of a teacher.)

Their social lives don't look very complicated either. Because they say what they mean, I find it easy to be with them. With my more Agreeable friends (like myself), I never really know where I stand. And vice versa. *I don't care where we eat, you decide. Oh, I don't care either…* People like me drive me nuts.

But low-Agreeable personalities pipe up. You can agree with them, or you can disagree, but you won't waste any time trying to decipher their true intention.

Super-low Agreeableness can cause problems. A person who invests every ounce of his effort in himself isn't so fun to be around. I recently got into a spat with such a personality. A minor elected official, this guy's politics tend to mirror his own personal interests. He is unmoved by anyone else's plight—until he experiences that plight himself. And then he blazes with outrage and self-righteousness.

And at the extreme end of low Agreeableness, stand the psychopaths. They're fascinating in what they tell us about the human condition.

The psychopath delivers a backdoor proof that we have some kind of mirror system in our brain. These personalities don't experience much negative emotion themselves. They have calm—eerily calm—amygdalas. The wiring that connects the amygdala to the rational PFC also shows signs of malformation. With his faulty fear equipment, a psychopath may neither notice nor care when he is rejected or insulted. As a result, he's severely handicapped in the reading of others' emotions. He's blind to your pain. He cannot detect it.

And what would each of us do differently if we believed no one would suffer from our actions? I imagine I would help myself to all the things I wanted, when I wanted them. Human morality presumably constitutes a biological aversion to harming others. If my behavior causes no harm, how can it be immoral? The typical psychopath does learn that certain behaviors cause other humans to cluster around and shout at him, and so he does develop some restraint. But this self-control is a cool, memorized version, not the warm, innate type that normal personalities incorporate.

Low and high Agreeableness represent two approaches to managing the conflicts that we social animals face. You can either appease all the animals around you, or you can march to your own tune and hope a few others follow along. And, of course, personalities fall all along the spectrum between the two extremes. Each person apportions her energy between herself and her fellow humans in a unique ratio.

4

FACTOR:
CONSCIENTIOUSNESS

CONSCIENTIOUSNESS FACETS
Self-Discipline
Orderliness

THE CAST

Mascot ....The Sugar Rat

Neurotransmitter ....Dopamine all over again

Brain Regions ....Striatum,
and the good old Prefrontal Cortex

CONSCIENTIOUSNESS

THE CONSCIENTIOUS PERSONALITY IS a combination of high morality and a long-term orientation. This person makes plans, anticipates problems, and marches toward her goals—one of which is to build a solid reputation among her comrades. At the lower end of Conscientiousness, a person remains open to changing course when a new opportunity pops up. This short-term orientation allows more flexibility. The low-Conscientiousness person is less constrained by a sense of obligation to the other people in his community.

Conscientiousness is a tricky thing to spot in rodents. We can see self-discipline in a rat, since some individuals are able to wait longer for a reward that's bigger. If they had no self-discipline, all the rats would take the immediate option. But some wait. These self-disciplined rats can control their impulses for a whole thirteen seconds.

The human animal, though, leads an extremely complex life that demands long-term planning. And long-term plans go nowhere without Conscientiousness. You can't build a house without a certain amount of confidence and commitment. Many days, building the house will be painful, exhausting, and otherwise aversive. But the human personality is equipped with features that allow us to plow through the pain as we pursue that long-term goal.

The human's complex lifestyle includes a complex social life. This, too, requires a lot of work. To build connections that will survive over the long haul, we have to commit ourselves to a program of social maintenance. We know hundreds of people, interact with thousands more, and rely on vast networks to keep us fed, sheltered, and healthy. We need these people, these networks. We can't get along without them. We cannot slink into the forest alone, like a bear or a tiger, and be perfectly self-sufficient. And so we have a built-in ability to sacrifice today in order to gain social support tomorrow.

People with high Conscientiousness are especially conscious of the future. Rather than wait for it to arrive, they plan, and go forward to meet it. They invest in their own future, they commit to their plans, and they dedicate themselves to their social circles.

People with low Conscientiousness invest less in the long term. They invest less in planning, and building, and sacrificing for friends. They direct that energy instead to the present. When opportunity knocks, they prick up their ears and they skip to the door. Because they're not already committed to seventeen other people and projects, they're able to welcome opportunity and make the most of it. They sacrifice stability, but they benefit from greater flexibility.

THIS GIVES YOU A QUICK look at where you land on this facet. If your answers tend toward the "often" side, you're higher in that facet.

Self-discipline describes an ability to forge ahead even when it's no fun. Commonly known as willpower, it's prominent in the personality who takes his body jogging every day, flosses his teeth, balances his checkbook, and calls in sick to work only on those days when he is actually ill. A personality with low self-discipline struggles to follow through on commitments, even when he has a genuine desire to succeed. This person is easily distracted, and in our goal-oriented culture may seem flaky and irresponsible.

Self-Disciplined Mouse

Marilyn Carroll has a black belt in discipline. She resisted a locker stuffed with marijuana for years. Even when friends wheedled—*just a little for the party this weekend*—she did not open the locker. She had inherited the weed from a previous occupant of her University of Minnesota lab, who used it for research. There were a hundred pounds of it. She hadn't a clue what to do with it, although she was adamant that smoking it was not an option. OK, maybe the temptation wasn't that great.

"The government grew it for research. It was terrible."

So the marijuana sat there, for years. Carroll didn't have time to research the proper disposal of musty recreational drugs. She was totally into pharmaceutical-grade cocaine.

She's a blonde with pale blue eyes, poised at the transition from pretty to elegant. But when she roams the halls of the laboratory in her running shoes she looks more like a teenage jock who has yet to hit the hurdles of arthritis and fraying tendons.

"We don't work with mice," she warns me. "We work with rats, a much cuter, less smelly, more highly evolved rodent."

Rats may be more evolved and less smelly (cuter is a stretch), but they're a lot harder to genetically modify. If you want to create self-disciplined rats, or impatient ones, you can't just turn off a gene. You have to breed them. Just the way you would develop a miniature pony from normal ponies, you select the most patient rat from one family and breed it with the most patient rat from another family. Over a few generations you develop a rat with legendary discipline—or lack thereof.

We head off to visit them. At a heavy lab door Carroll swipes her card and kicks the balking steel in a cadence that suggests long familiarity. The doors are a hassle, but so are animal activists. Carroll works with monkeys in addition to rats (monkeys are superevolved, smelly, and still not as cute as Mitzi and Maxi, my desk mice). This rouses the passions of animal rights types. Carroll recalls how years ago a swarm of activists stormed her office and lay down around her in an arc. They linked their necks together with bicycle locks. The kookiness was compounded when a campus policeman arrived with a hacksaw.

"I said, 'You're going to cut his jugular vein. He's going to bleed out on my floor.' He said, 'OK, I'll get some paper towels.'

"I left at that point."

We turn into a room where white cabinets line the walls. Inside each cabinet, sheltered from light and noise, is an aluminum pagoda with a plastic door. These are experiment chambers where rats come to press bars and demonstrate their patience or impulsivity. Carroll opens a cabinet and studies the occupant. Circling the perimeter of the pagoda is a wild-looking white rat.

"She's pretty jumpy," Carroll observes. "I think she's had some co-caine." She checks a syringe on the side. "Yup." A fine tube runs from the syringe to a "backpack" on the rat, and from there into an artery.

The rat is just a normal lab rat, not inbred for a particular response to drugs. But this is where Nature meets Nurture. Carroll studies (among other things) whether an environment that includes addictive drugs af-fects every personality the same way. And what effects drugs have on a personality over time. Particularly, it would be useful to know if the "nur-ture" of drugs erodes natural self-discipline, causing drug use to climb.

To find a rat's baseline of patience, Carroll subjects them to a choice: When a light goes on in their pagoda they can push a bar immediately and receive a single yummy food pellet, or they can wait a few seconds and get three. Rats are smart enough to grasp this concept. And once they get it, Carroll can stretch the delay time to pinpoint each rat's degree of self-control. A self-disciplined rat can wait thirteen seconds. An impa-tient rat goes for the immediate reward.

After the rats have shown their colors Carroll can test them against other temptations, to see how different personality types will react: How will natural self-discipline interact with the addictive quality of cocaine? Does it help a rat avoid addiction?

In a word, yes. A rat who can wait for a bigger reward can also resist cocaine. Likewise, he is hardened to the allure of nicotine, alcohol, sugar, and excessive exercise.

To determine this, however, Carroll first has to hook the rats on drugs. Hence the thin tube running from the backpack to a syringe outside the pagoda. For six hours a day a rat will visit the pagoda to receive periodic infusions of cocaine. Her dopamine system takes note. It begins to want the drug. *More . . .*

Fifteen seconds before each infusion a lever slides into the pagoda. By pushing it, a rat can get that infusion immediately instead of waiting. By the fourth week of training most of the impulsive rats figure that out. Their dopamine systems have locked on to the target. *MORE.*

The next question is, does your personality influence how much drug you want? Now the lever stays in the pagoda, and the light goes on one hundred times a day signaling that a dose is available. Many of the im-pulsive rats will smack the bar to collect every dose. Impulsive females are especially quick to reach the hundred-dose limit, which they'll continue to use day after day. If you allowed them ten more infusions they would take those, too. Impulsive males are slightly harder to hook—only about two-thirds of them max out. Even so, they're much better customers than the self-disciplined rats. That group just isn't into drugs. They are much harder to hook. After a full month only about 10 percent of the self-disciplined males are hitting the cocaine a hundred times a day; the self-disciplined females are a bit more susceptible. Cocaine simply doesn't rouse the dopamine system of a self-disciplined rat the way it does an impatient one.

Not long ago scientists believed that an anxious personality represented the highest risk for addiction. Carroll and others are calling attention to a different part of the personality. The impulsive, impatient personality now demands equal respect.

Although the rats in this lab are exercising their impulsivity on cocaine, the personality type is prone to other addictions as well. Carroll is particularly interested in how impulsivity might affect a person's ability to resist food. Obesity is becoming such a common and expensive problem that food addiction is starting to look just as dangerous as addiction to meth, crack, and other stalwarts. Carroll has long suspected that drug addiction and overeating are linked.

She began working with what I call "sugar rats" because they are such model bingers. Scientists bred the line by mating together individuals who loved the sweet taste of sugar or saccharine.

"When these rats binge, they really binge," she says. "A rat normally drinks about twenty-five milliliters a day. But the high-saccharine rats will drink five times as much. That's about half their body weight." Calculator, please…for a 150-pound person that would be more than nine gallons of diet soda! Per day!

And is a rat who binges on sugar or saccharine quick to acquire other addictions? Yes. If you put two bottles in a rat condo, one with water and the other with water-plus-alcohol, the sugar rats will hit the spiked bottle. In fact, the stronger you mix their drink, the better they like it. Sugar rats have also demonstrated their aptitude as cokeheads and heroin junkies. I find the sugar connection piquing because I've seen it in action. I've known a handful of alcoholics who, when they put down the bottle, picked up the ice cream.

Just as intriguing, Carroll has studied rats who become addicted to running on a wheel. As with self-discipline, she found that a roomful of regular rats will show tremendous variation: Some rats run four times more than others when they're provided with a wheel. Some rats will run instead of eating, and lose weight. As with sweets and cocaine, female rats are more likely than males to go to extremes.

Once again, Carroll sorted out the marathon-running rats, to see how this personality type responds to drugs. Whaddya know? The same rats that love to run also become easily addicted to cocaine.

So, the bad news first: An addictive personality is vulnerable to a grab bag of addictions. The good news—great news, really—is that the temptations are somewhat interchangeable. That's true even after addiction has sunk its teeth into the brain. Carroll has shown that a rat who's jonesing for a dose of cocaine will settle down if she gets a dose of saccharine instead.

Now, any monkey can see that human drug addicts are harder to convert to an ice-cream habit. Nonetheless, some of the most promising drug-treatment programs are now exploiting this flexibility in the dopamine system, using vouchers to reward people who can fight off their drug cravings. Addicts who participate can use the vouchers for healthy substitutes for a drug high. These range from fabulous food at a restaurant, to distraction at a movie theater, televisions and other gizmos, and in one research project, even jobs. These incentives help to mute the clanging of a disappointed dopamine system.

Carroll kicks another door. Those sneakers are more than a fashion statement.

We're in the people room now, where young scientists collect equipment or rattle away at computer keyboards. On a desk sits a plastic rat condo with two inhabitants. "These guys are pets," Carroll says, peeking through the roof. "You look like you live in a Dumpster." Shredded paper towel, hunks of printed cracker boxes, and more colorful debris litters the condo. The occupants peek back, sniffing. I let one toy with my fingernail, which is too broad to get his teeth around. Except that he does. Fighting my impulse to jump and squeak, I lower my hand. I dab blood on my black pants but I'm bleeding so fast this won't do. I feel like an idiot. Sidling into a supply room, I twist paper towel and masking tape around the wound. I'm caught. Carroll tries to impugn the rat's character, not my own.

Anyway, now we know which personality type tips toward addiction. But what does that personality look like, inside the brain?

Most research these days concentrates on the dopamine system, the same neurotransmitter we encountered in Extraversion. Clearly, dopamine is also involved in addiction. But where? And how? Before you create a drug or tailor a therapy, it helps to know what you're fixing.

Traditionally, an advantage of studying rodents is that you can investigate their brains to analyze exactly how the nerve cells and chemicals are interacting. That often results in the demise of the rodent. Ah, but that was before rat PET. Yes, Positron Emission Tomography, now in rat size! This is like an X-ray, but with a custom tracer added to the rat's (or your) body to highlight the cells you're interested in. A sedated rat glides into the miniature scanner, reveals his brain, and glides back out hale and hearty. MRI machines, too, have shrunk to rat and mouse sizes.

The magic of micro-imaging is that now you can capture portraits of a rat's brain before and after you give it drugs, to see if anything changes. Is there something different about an impulsive rat's brain, even before it encounters cocaine? Or does cocaine enter a normal brain and alter it? Does an unusual dopamine system predate the drug use? Or does cocaine change the function of the dopamine system?

In this vein, an English team with a research mission akin to Carroll's recently scanned a dozen rats—six self-disciplined, six impulsive—using rat PET. None of these rats had ever experimented with drugs, so the team was snapping the "before" portrait.

And they did find a difference. In the impulsive rats, they found a dearth of dopamine receptors—the locks of the system—in the lower part of the striatum. The striatum is one of the middle "onion layers" of the brain. Its upper side is related to learning. And that region looked normal, which you would expect in an animal who quickly learns how to score a drug fix. But when researchers counted the dopamine locks on the underside of the striatum, they came up short.

That region is known to flare with excitement when an animal stumbles onto one of those life-giving things like food or a receptive mate. It's more instinctive than the upper side. It doesn't need to learn what it likes. It just knows. Dopamine tells it.

This shortage of dopamine receptors supports the ascendant theory of addiction: People (and rats) don't get hooked because their dopamine reward system is overactive. They get hooked because it's not very sensitive. It needs a big bonk on the head to feel satisfied. Denied that satisfaction, it just keeps moaning for more, *more*, *MORE*.

In the English experiment, the scientists wanted to be sure it was dopamine receptors, not dopamine itself, that was in short supply. So they also fitted the rats with the kind of beanies I saw at the Neumann lab in Germany. Microdialysis tubes descended into the brain to monitor chemical levels in real time. And they found that the rats' dopamine levels were normal. The rats had enough keys to activate a normal number of locks. And therefore it is likely that addiction is aided by a shortage of locks, not a shortage of dopamine.

The end result, of course, is the same: In these brains, addiction is an accident waiting to happen. It's sometimes tempting to blame addicts for their plight. We who have a normal number of receptors in our lower striatum just cannot imagine why a person would behave so slavishly with a drug. Likewise we look at people who weigh three hundred pounds and think they lack strength of character. We cannot imagine how it feels to lack strength of dopamine.

But do we blame a rat for becoming easily addicted to cocaine or a running wheel? That would be pointless. Among rats, addicts are born, not made. Certainly you can turn a self-disciplined rat into an addict if you train her dopamine system rigorously. But judging from Carroll's bar-pressing experiments, self-disciplined rats are never going to be better than middling drug customers. They just don't have the drive.

In another room a couple of rats recover from surgery. Carroll picks up a patient, who drapes like a white sock across her palm. She pats the soft back.

"In graduate school I wanted to be a social psychologist. I didn't want to work with rats. I thought they were disgusting. But I got into eating issues with rodents. And before long I loved it."

She lays the rat gently down. I'm experiencing a twinge of rat fancy. I adore my Mitzi and Maxi, but they're not very huggable. A mouse is just too small to snuggle. But a rat—that's approaching a squeezable, if not huggable, size.

Self-Disciplined Human

Carroll walks me across campus to the lab of a colleague who studies self-discipline in humans. A polar wind drives bits of ice into our cheeks. Minneapolis is a very northern place. Northern and outdoorsy. Carroll lives well north of the city, among frozen lakes. The ringtone on her cell phone is the yodeling of a loon. We pass very few people, and each of them is hunched, bundled, flapping.

Inside, the research assistant Stephen Duquette sits me down to test my self-discipline. The room is not what I expected. I don't know what I did expect, but it wasn't a closet with a vanilla paint job. The narrow room lacks windows and art, anything that would distract people. I sit at a computer terminal.

Duquette hands me a box with two buttons to push. This is how I'll record my responses. The first test is the human version of the rat test that offers one food pellet now, or three after a delay of thirteen seconds. I don't want to tarnish my reputation for modesty, but I may have set a new record for dullness and lack of spontaneity.

"Would you rather have $1 now or $100 in two days?" the computer asks.

Duh. But the test is just warming up. It moves from obvious contrasts to closer calls.

"Would you rather have $15 now, or $17 in a week?"

"Would you rather have $10 now or $15 in a year?"

I feel intuition jerking my brain this way and that, and I struggle to do the math. I mean, waiting a year to gain five bucks may sound silly, but it's a 50 percent return on investment. That's not silly at all.

It's exhausting, and it lasts perhaps ten minutes. Then Duquette calls up a graph of my results. Typically, a person's self-discipline is lower the longer you ask her to wait for a payoff. So most graphs slope down as the wait grows. My graph is a straight line that hugs the top of the graph.

"That's not much of a curve," muses Duquette. "You're willing to wait." It's true. I'm that investor who bought stocks ten years ago and although they're currently worth half what I paid, I will not sell them at a loss. I'm willing to wait.

That's one test in the arsenal. There are more. They're worse. Three "X" signs appear on the screen in a row: X X X. Then they vanish. Another row appears: X X X. And then a different row: X O X. I'm supposed to click a button when that happens. I begin to press the button, but suddenly a little frame appears around the O. I'm not supposed to click the button when that happens. Can I stop myself? This measures a different side of self-discipline, the ability to inhibit your impulses.

It's a horrible test. Now I feel a different strain in my brain, as I try to predict what will happen next. But it's unpredictable. It's my physical impulsivity, and my brain's ability to halt that impulse after it has started, that Duquette is measuring. I tense for a row of symbols. Between the crosses an O shows its face. My click muscles fire, but then the %#*& box appears around the O. Too late. I clicked. I feel like a failure. Awful test.

The Stroop test is next. A classic that's easily found on the Internet, it presents the word "purple" in red ink. And the word "green" in yellow ink. And "blue" in purple ink. And so on. You name the color, not the word. When the human brain sees a word, it is so primed for reading that it's very hard to override that initial impulse. The brain wants to scream out the word, not the color. You can feel your neurons bleeding as you stare at the word "purple" and try to say "red."

There are more tests of self-control and impulsivity, too, but I decline. Even though I'm performing like the world champion of joyless obligation, I still feel like a giant loser each time I click a framed O. I suppose my high Conscientiousness tells me I could always do better. Because I am an unparalleled drudge, it's a struggle, but I do it: I push back my chair, and I quit.

Of course there are no winners or losers in personality. There is just more this way, or more that way. A self-disciplined personality has more control over the impulses that course through everyone's brain. Every personality type is lured by temptation. Without it, we'd go extinct from starvation and failure to breed. But the self-disciplined personality is able to squelch some of those impulses so that she can pursue long-term goals.

The risks associated with a highly disciplined personality are...uh... there are none. Well, that's not true. For the most part, our culture embraces long-term goals. We're builders—of houses, careers, families, reputations. All this building necessitates planning, patience, and persistence. So self-discipline is a nice characteristic to have. But it confers certain limitations. Because I'm building all this stuff, I can't very well spree off to spend a few years on Corsica, or even walk away from my desk today to go kayaking. Because I opt for long-term investments, I forgo the short-term thrills of playing the lottery and flying to Vegas for a wild weekend. Because I don't bet big, I will never win big. That's one cost of a self-disciplined personality.

There is also a social cost. Just ask my stepson which type he prefers in a parent. His father is the deliverer of donuts, the forgetter of rules, the patron saint of indulgences. I, by contrast, am a nag. And when I consult my watch and announce the hour of homework, vitamins, tooth brushing, bed, laundry, vegetables, dog-walking, milk, fresh air, or fish oil, four eyeballs roll. It's not fun, for any of us. It's just who I am: not fun.

I've had the chance to study the alternative, however, and I'll stick with my end of the spectrum. Spontaneity and impulsivity have their place. They have a lot of places, in fact, and can be really fun. But in the modern environment, fraught with temptations, this personality is at a terrible risk. That dark side of impulsivity is the whole reason Marilyn Carroll has a big lab full of researchers and rats.

My husband is a sugar rat, born and bred. He was in his teens when he first tried drugs. That's a typical pattern for the human sugar rat: He encounters a new substance or experience, and he pushes the bar. He doesn't pause to reflect on what his parents would advise. He doesn't analyze how being wasted might affect his homework later that evening. He doesn't produce a forecast on his ability to drive home. He pushes the bar. And in our current environment, drugs and alcohol are so common they might as well be lying on the sidewalks.

Like a sugar rat, my husband quickly found himself pushing the bar every day.

This is where brain science fades out. We still don't know how addiction alters the brain in such a predictable way, but clearly it does. The disease of addiction follows a pattern that often ends in the death of the animal, whether rat or man.

During the acquisition phase, the brain develops an obsession. Some animals acquire an addiction easily—my sugar rat was a natural. Others have no talent for it. I experimented with various drugs, but to my brain it all seemed a lot more trouble, and money, than it was worth.

Once a brain is hooked, it settles into a state of maintenance at a certain level. The maintenance dose depends partly on how hard the animal must work for a fix. Rats vary widely in this, with some willing to press a bar hundreds of times to earn an infusion, while others are less motivated. And the same goes for humans. Some people will discover that their craving pushes them into a life of stealing, while others can level off when they're spending a chunk of the family budget so modest it goes unnoticed.

But escalation can raise the maintenance level at any time. Sometimes a binge on cocaine can promote an addict to the next level, permanently. Now he needs more than he can squeeze out of the grocery budget. Some drugs cause escalation by their very nature, as the brain builds up a tolerance to them. The dose that soothed the dopamine system yesterday isn't able to do the trick today. The craving pulses harder. *MORE. MORE. MORE.*

My sugar rat's addiction escalated in fits and hiccups through his twenties. If a bar slid into his pagoda, he pushed it. He didn't contemplate the cost. He didn't imagine the rewards that could result from abstaining. That's not who he is. He acts. And once the dopamine system locks on to an unhealthy target, its demands are usually going to escalate.

Addiction is a brain disease. As it progresses, it causes an animal to abandon the natural animal behaviors that promote life and health. Rats addicted to cocaine turn their back on food and fellow rats. They stop cleaning their fur, lose weight, and ignore the bottle of sugar water that once delighted them. In one experiment, 90 percent of rats with unlimited access to cocaine were dead in a month. In the progression of my own sugar rat's disease, cars crashed, surgeons hovered, jobs imploded, parents despaired.

My sugar rat was fortunate. By the time he was thirty his rational prefrontal cortex had deciphered the writing on the wall. This couldn't last. He was a medical man, a man of science. He acknowledged that his dopamine system had been corrupted, and that only hard work could salvage the situation. And it is hard work. Once derailed, the dopamine system remains off course for life. My sugar rat will never again be unaddicted. Day after day he faces down the dopamine with knowledge and sheer willpower. Almost every day he sits down with people in similar circumstances to practice managing their dopamine. Learning from each other, they train their brains to overpower dangerous impulses.

Like rats, humans can become addicted even if they're not wildly impulsive. Different personality types have different relationships with those addictive substances. My sugar rat is strongly extraverted, a personality type associated with a dulled dopamine system. That type gravitates toward stimulant drugs such as cocaine and "speed." Presumably, an extravert's dopamine system is responsible for both inclinations: Social interaction and stimulant drugs both appease a chronically needy brain.

My father, by contrast, was an introvert and powerfully anxious. His

drugs of choice were alcohol and tobacco, which are typical of the Neurotic personality type. While all addictive drugs activate the dopamine
system, alcohol also has an intense relationship with the amygdala.

My father was able to withdraw from alcohol without treatment or
support after he realized his friends were avoiding him. (He became a
sugar junkie.) But his dopamine system was so clenched in nicotine's grip
that he tore free only months in advance of a scheduled surgery. He, too,
was a man of medical science. He knew the lungs are the weak link in an
anesthetized body. Fear gave him the strength to break with nicotine, al-
beit too late. To support its habit, his brain had secretly sold his lungs
down the river.

So the anxiety facet of personality is just as effective as impulsivity,
when it comes to pushing a brain toward addiction. And of course neither
personality facet is a guarantee that you'll latch on to the first drug that
comes your way. But they sure help. And if you are ever given a choice,
anxiety might actually be the preferable path. At least in that case you
might stand a chance of having some self-discipline. And that clearly
helps in the recovery phase of addiction. Relapse is terribly common-
place for all addicts, but personalities with high impulsivity are the most
vulnerable.

As Marilyn Carroll suspected all those years ago, low self-discipline
leads as easily to overeating as to overdrinking or overdrugging. And
we're seeing that the result of food addiction can be just as destructive. In
many cultures where surplus food has flooded the landscape in recent
decades, obesity is becoming an epidemic disease.

Like most addictions, it's the side effects that end up killing you. Alcoholism wears out the liver and the gut. Methamphetamine erodes the
teeth and the brain. Nicotine poisons the lungs. And food addiction ex-
hausts the pancreas and the circulatory system. This addiction, like many
others, can be fatal.

The brain biology is much the same. People who overeat share the
same dopamine profile as other addicted people. The gene that builds
dopamine receptors—locks—looks increasingly like a misfit with the
modern environment.

Actually, it's only one version of the DRD2 gene that's turning up like
a bad penny in study after study. Once upon a time, the A1 allele of
DRD2 may have motivated a human to charge forward in search of new
flora and fauna. A1 increases your odds of having the dopamine style of

a sugar rat: hard to satisfy, always wanting *MORE*. But back then, the human environment didn't offer so many opportunities for addiction.

People whose dopamine system is on the demanding side have to watch their step in an environment fraught with pleasures. Their dopamine receptors are scanty, so even when a wave of dopamine keys is breaking over them, the system may not activate normally. Dopamine molecules, the keys, pour into the brain's intersection in good supply. There just aren't many locks for them to activate. So the dopamine system keeps banging on a bell—a really loud, harsh, painful bell: *MORE! MORE! MORE!*

It's not fun. The craving that addicts feel is not the anticipation of a joyous experience, research now suggests. It's a yearning for the noise of the dopamine bell to stop. It's the craving for an end to gritty discomfort. Your average addict, whether hooked on nicotine, alcohol, caffeine, cocaine, ice cream, sex, or lottery tickets, does not feel like a million bucks when she finally feeds the dopamine system what it's been clanging for. She might feel euphoria and relief for a little while. But shortly thereafter she feels, typically, like a failure.

And then, of course, there are personalities like Marilyn Carroll. She's schoolgirl trim. There is no trash can full of coffee cups in her office. And she had a hundred pounds of pot in a locker for years and never smoked so much as a joint.

"I eventually called the DEA and asked what I should do with it. They told me to put it in a Dumpster. If I thought that would be a problem, I could burn it in the parking lot."

She rolls her eyes. Granted, the legal conundrums faced by a drug researcher who experiments on animals are likely to be a tad unusual. But between this drug-disposal guidance, and the campus cop with his hacksaw and jugular veins and paper towels, Carroll's experience with law enforcement has been batty. Loony. Ratty.

"I did finally put it in a Dumpster on a Friday night," she sighs. "I hope I gave someone a nice weekend."

The question of whether drug use erodes self-discipline over time remains unresolved. Research in both rats and humans yields hints, but not bold-faced answers. It's worrisome that many studies find both animals become even more impulsive when they're regularly using cocaine. But other studies, some addressing other drugs, find that a personality returns to its baseline of impulsivity after a period of drug-free living. The truth is written so faintly in the data that nobody is ready to stake her reputation one way or the other.

And that is probably good news. If cocaine and other drugs permanently savaged the self-discipline factor of personality, then studies wouldn't be so debatable. The results would leap out in big numbers that said: *Drugs ruin your self-control forever!* The fact that the numbers are barely whispering suggests that the effect is temporary.

Evolution of Self-Discipline

Why has the human race retained a personality type whose dopamine system can lead a person so directly to disaster? What is so beneficial about that? How does addictive behavior promote the survival of an animal, whether mouse or man?

This seeming absurdity makes more sense after we blow away the excesses of our modern environment. Let's brush away the fancy drugs—cocaine, meth, heroin, nicotine, caffeine, and the like. Until recently these chemicals were derived from wild plants. They were weak, and not widely known. Tobacco was used only by American Indians. The coca plant was known only to natives of the South American Andes, who chewed the leaves for a mild energy boost. Before global trade, most people had no access to such drugs.

Next let's eliminate the processed foods. Even a century ago cookies were a rarity, demanding expensive sugar and coveted butter. A century ago most humans on earth harbored intestinal worms that competed for the lean, whole-grain foods they were able to acquire. For the average person, something as rich and effortless to digest as a bagel with cream cheese was beyond the imagination. So let's be rid of bagels, cookies, white bread, white rice, pasta, and all sweets and pastry.

Alcohol has been with us much longer, perhaps twelve thousand years in some cultures. The earliest attempts were beerlike brews made from fermented grain. Anything more sophisticated would have required a settled lifestyle. Like natural drugs, these would have been light beverages delivering more of a nudge than a kick. But even twelve thousand years isn't sufficient time for evolution to weed out the weak dopamine genes from the human genome and harden our brains to alcohol's appeal. So it's gotta go.

Salt must also be puffed off the table. Coastal people have always had it, but before industrial mining, it was a coveted trade item. Those who lived near a natural deposit were in luck. Those who didn't could get some by eating meat, and a little more by burning plants and eating the ash. Blow away the salt. Until recent centuries, it has been hard to find.

Survey this new landscape, an environment with no white flour, sugar, salt, alcohol, coffee, alcohol. Survey this sugar-free landscape through the demanding eyes of your dopamine system . . . nothing. There is nothing here for dopamine to get excited about. Maybe there's a prickly plant somewhere that you could chew for a gentle buzz, but it takes a lot of chewing to get a little lift.

Even an animal who's born to be addicted has a "break point." That's the point where an animal decides that getting a fix is more work than it's worth. And a few centuries ago, daily life was one enormous, wet blanket of a break point. Anything you could get hooked on was too expensive, too weak, and demanded too much chewing. Under that scenario, what does it matter whether your dopamine system is tyrannical? It can clang and bang all day, but there's nothing in the environment for it to fixate upon.

Go back deeper in our history, back to the forests and plains of our species' evolution. Go further still, until you encounter the ancient ancestor of both rats and you. Here we'll see the original benefit of a tyrannical dopamine system. Dopamine's role in pushing animals toward the things they need is ancient. And quite simple.

Your basic animal is a tube. At one end is a hole that vacuums up food. In the middle the food seeps into the animal, fueling it for future acts of ingestion and reproduction. At the far end is a second hole for ejecting waste. Often a second opening near the waste hole facilitates reproduction. That's the basic design.

Propelled by limbs, wings, or fins, this tube moves through the world in a food-seeking orientation, mouth first. When it locates food it ingests it and moves on. What tells it to move on? To find more food?

Dopamine. Without dopamine an animal would sit like a stone, wanting nothing. In fact, rats under the influence of a dopamine-blocking drug lose the motivation to eat. When researchers put food into their mouths, the rats chew and swallow it eagerly. They like it. They just don't want it. Obviously such an animal would not survive in nature. Equally obvious, evolution would therefore prevent such a system from existing.

Evolution will also leave behind an animal who wants things, but indiscriminately. Life takes calories. Life takes salt. An animal who craves pebbles and snowflakes and beetle wings will never save up enough energy to make offspring. That diet is a nonstarter.

So a successful mammal, like the mouse or the human, enters this world with her dopamine system already tuned to the foods that have been crucial to our ancestors' progress. From day one, my dopamine system pointed me at salt, a mineral so hard to find and yet so crucial. It pointed me at high-calorie foods. Etched in my dopamine system is the pattern of eating that guided my forebears safely through millions of years of foraging.

Sure, my great-great-et-cetera-grandmother could have met her caloric needs by gathering low-calorie leaves and insects. But that's inefficient. Every hour she spent working far from safe shelter was another hour that the lions could track her down. Starchy roots and sugary fruits have more calories than leaves do. If, guided by the demands of her dopamine she could fill her calorie quota in half the time, then her risk of being eaten was cut in half, too. Her chance of surviving that day doubled.

So that's why we have dopamine. Evolution has made it effortless for us to orient our food-intake hole toward potatoes, not pebbles. With dopamine as our compass, we home in on the most efficient and essential foods. In other words, we don't eat French fries because they taste good. They taste good because we need to eat them.

We used to need such rich foods, anyway. Then the environment changed. It happened suddenly and radically. In the developed world we spend our days surrounded by high-calorie foods. The Skor bar I just ate gave me 200 calories in two minutes. I'd have to chew my way through four apples to match that. A mid-size potato has about 150 calories, while the same weight in French-fried form delivers four times the fuel. Today, instead of hydrating with water like other animals, a human can wet the whistle with beer or soda or wine and gain 150 calories. And salt is everywhere. Even in the Skor bar. In this environment, we don't need to be spurred toward food. Food practically jumps into our mouths.

The environment has changed, but our dopamine system hasn't caught up. It's troubling enough that dopamine drives us to consume more than we need. What's worse is that it remains willing to adopt new targets throughout our lifespan. We come preloaded with a program to seek out food, water, sex, and other essentials. But we're also able to add to the shopping list at any time.

Why would evolution favor a system that can be programmed after birth? How does it benefit an animal to develop an obsession?

I suspect it has to do with the fact that environments change, and humans wander. So perhaps we're born with a dopamine compass that points us toward the basics, but can also be programmed to suit new environments. Like a bloodhound who can be primed to seek lost children, or drugs, or bombs, perhaps our dopamine system evolved an ability to change targets on the fly.

Perhaps humans start life with a simple food drive: *More calories.* But as we grow and eat our way through our environment, the dopamine system becomes more specific, pointing our mouths at the most efficient food sources this place has to offer. In one ecosystem, that may be starchy roots or tubers. And the dopamine system refines itself to automate the searching process: *More carbs.*

In a different environment nut trees may shade out the ground plants. The dopamine system slowly registers this and adjusts its message: *More fat.* Or if fruit is the quickest meal: *More sugar.*

The dopamine system is mutable because the environment is, as well. That's my hypothesis. That, and $2.25, will get you a cup of coffee.

But speaking of coffee, the mutability of the dopamine system leaves it vulnerable to hackers, including caffeine. Caffeine delivers no calories, no essential minerals or vitamins, nothing of any use at all. But it fools the dopamine system. And after a few doses, the dopamine system adds caffeine to the list: *More coffee.*

And so it goes with alcohol, nicotine, cocaine, and other addictive drugs. They're useless, in terms of furthering an animal's caloric and reproductive goals. But in the human brain, they act like dopamine keys. And the dopamine system automatically adds them to the list, too: *More carbs. More salt. More cocaine.*

This is a major malfunction, in today's environment. It would be OK to be impressionable in a poorer environment. But to remain open to suggestion in a world that's awash with suggestions is perilous.

Fortunately, most of us have a dopamine system that doesn't sign on at the first taste. Most people are like Carroll's rats who are offered a hundred doses of cocaine but don't push the lever very often. You can build an addiction in them if you apply a drug consistently, over time. But they're not naturally gifted addicts.

The *carpe diem* people make your best addicts.

We met them back in the chapter on Extraversion. Impulsivity is a major feature of that outward-oriented factor. And at the extreme end of that personality spectrum we found the superimpulsive personality that's now called ADHD. These personalities are also at a super risk for addiction in our substance-rich world.

I would venture to guess that the addictive personality persists in the gene pool for the same reason that the impulsive one does: It's a short-term orientation, ideal for an unpredictable environment. If your environment goes through boom and bust cycles, or drought and flood cycles, or no-acorn and many-acorn cycles, you can benefit from a flexible strategy. Likewise, if you're inclined to wander, you can easily find yourself in an environment unlike any you've seen before. You'll probably do better if you enjoy exploring the new possibilities instead of pining for the habitat you left behind. And if you grab at those new possibilities without pausing to ponder the long-term ramifications, you might stand a better chance of making it through that day. There is much to be said for seizing the day, for the bird in the hand, for the stitch in time.

Self-discipline has its advantages as well. It's not the most flexible approach to life. But especially in a stable environment, waiting often pays better than grabbing. With time, fruit gains sugar (calories). With time, animals (protein) lose their fear, or forget you're nearby. Humans learn to trust you, with time. Fish circulate into the shallows for an easier shot, with time. Many foods are safer and easier to digest if you take the time to let them cook. The self-disciplined animal sacrifices immediate gratification for a bigger reward later. This pays handsomely in quiet times.

One of the curiosities of Carroll's rats is that they don't seem to come in "moderate." While we reverse our trek across the university tundra, she talks about her initial study of self-discipline in normal rats. "In the early screening we looked at a hundred and sixty rats. We gave them a delay discounting task. We screened all these rats. Most of them were under eight seconds; a handful were thirteen seconds; and there were almost none in between. Of all those rats, only two were in between."

Apparently moderation in self-discipline is not a useful quality in a rat. For a rat, at least, being sort-of disciplined, or halfheartedly impulsive, is a formula for failure.

Conscientiousness Facet:

ORDERLINESS

ORDERLINESS INDICATORS	RARELY	SOMETIMES	OFTEN
I PUT THINGS AWAY WHEN I'M DONE WITH THEM	❏	❏	❏
I'M DETAIL ORIENTED	❏	❏	❏
PLAN YOUR WORK, AND WORK YOUR PLAN	❏	❏	❏

THIS GIVES YOU A QUICK look at where you land on this facet. If your answers tend toward the "often" side, you're higher in that facet.

The personality with high orderliness is calibrated for long-term success. This personality builds from the ground up and doesn't cut corners. When her ship comes in, it has a safe place to tie up, a crew on hand to unload it, and the dock is nicely swept. Of course, if her ship doesn't come in, she has wasted a great deal of time. Low orderliness—well, we all know what that looks like. This personality lent all his rope to a friend, and his dock is falling down. He's not repairing it because it still kinda works, and he's been spending time at this cool shack on the beach—hey, is that a chest of pirate treasure glinting in the sand? Low orderliness keeps a personality free to jump on short-term opportunities.

Orderly Mouse

Uh ... no. This is not a behavioral dimension that we can identify in mice. The single, solitary observation I can muster is this, gleaned in the mouse rooms at the Lesch Lab in Würzburg, Germany:

BalbCs, the white lab mice, arrange their paper-towel shreds in one tidy igloo. The igloo always sits in one corner of the condo. All the mice share it.

By contrast, the Black 6 mice shred the towel but then can't agree on where to arrange the shreds. Half an igloo rises in one corner, and another half starts in mid-condo, but it soon bumps against the descending kibble rack and the project is aborted. Shreds lie throughout the cage in various stages of being stolen from one igloo or the other.

Orderliness? That's a stretch. Nesting instinct degraded by decades of domestication? More likely.

Orderly Human

But humans certainly come in a range of personalities from orderly to not-so-much. This facet of personality looks like a close relative of anxiety, a facet of the Neuroticism factor.

Case in point is the "orderliness disorder." Obsessive Compulsive Disorder (OCD) is an anxiety disorder we first explored in the Neuroticism chapter, along with panic disorder, phobias, Post-Traumatic Stress Disorder, and other anxiety disorders. The most common OCD behavior involves contamination—specifically, avoiding it.

Here we find the compulsive washers who might scrub their hands thirty times before eating. With them are the people who fear germs, blood, saliva, or other bodily components, and those who loathe walking on the dirty ground or touching surfaces they consider contaminated. And they use ordered, methodical behavior to reduce their fear.

The second most popular category of OCD behaviors involves symmetry. A friend of mine feels uncomfortable and out of balance if she is touched on one side of her body but not the other. Her mother used to laugh about how Jill would try subtly to manipulate you into brushing her left hand if you had touched her right. Her mother was laughing with, not at. Her mother was my friend "Christina," from the Neuroticism chapter, who would drive home four times to make sure the oven was off. One day a bunch of us got talking about these quirks and discovered that M&M candy is an underappreciated test of OCD. One by one, we all confessed our M&M rituals. I vastly prefer to eat them in even numbers, and this number should be equally divided between my left and right teeth: three to the left, three to the right. The color should be coordinated at least (red, orange, and yellow can go together; green and blue can go together; and brown ought to be alone). But pure color segregation is ideal.

Many personalities, especially if they lean toward the anxious side, have a touch of order disorder. My grandfather couldn't rest at the table if his silverware was out of alignment. Millimeter by millimeter he would nudge them into perfect relation to one another. He was a humorless and rigid Yankee, so I found nothing endearing in his compulsions. But often our little fetishes are pretty cute. Remembering that when I was a toddler my older sister had warned me I had to practice doing everything with both hands so I wouldn't get lopsided, I asked if she retained any such compulsions today.

"Well," she replied, "I went through a phase where in order to go to sleep, I had to encircle myself with all of my stuffed animals, and then sequentially kiss each one a certain number of times. I think it got up to some ridiculous number like fifty or one hundred each! Does that count?" These days, though, she eats M&Ms by the handful, with callous indifference to color or number.

What contamination and symmetry obsessions have in common is that they bring order to a disordered world. Certainly germs are everywhere, and the fewer we ingest, the healthier we are. And symmetry is by definition orderly, predictable.

We have a second reason to find symmetry compelling, too. Humans and many other animals make a subconscious survey of the symmetry we see in the faces and bodies (and stripes and tail feathers) of animals we meet. And generally, when biologists offer animals a choice of mate, the more symmetrical candidates are more popular. Humans, when asked to choose the most attractive face from a stack of photos, are likely to select the most symmetrical face. One study has even shown that women can subconsciously detect symmetry in men just by watching them dance.

What's so obsession-worthy about symmetry? Well, the entire cell-division process that builds new life in the womb is supposed to be symmetrical, producing left and right sides of a person that are identical. But it never works perfectly. Every living thing turns out crooked, some more than others. And that degree of crookedness matters. Generally, research suggests that the most symmetrical animals are better, stronger, smarter.

So perhaps the OCD fixation on symmetry relates to our preference: These people's brains are on high alert (too high alert) for dangerous asymmetry. That's another of my coffee-shop hypotheses, anyway.

And as a symbol of my commitment to this hypothesis, I'm going to propose that the entire facet of orderliness, along with OCD, belongs under the Neuroticism tent, not here under Conscientiousness. But here is where psychologists place it (for now), and here I'll leave it.

Evolution of Orderliness

A dose of orderliness has many uses, especially for an animal whose tool kit includes a variety of lethal weapons. From fire to poison for arrows, and the arrows themselves, an early human campsite would have been bristling with things to trip over, fall into, and accidentally ingest. Human babies come into the world with no instinct for avoiding any of these hazards.

With her genetic future toddling around and grabbing everything in sight, a mother who always put her butchering tools away would probably lose fewer kids to household accidents. She who stored a few days' food against bad weather would stay stronger. She who planned her work and worked her plan would get through each day with greater efficiency.

And until humans were able to store a food surplus, efficiency was just as important to a human as to a mouse. Wasted time and wasted effort were a waste of calories, and every calorie was collected at considerable risk to health and welfare.

So a person with high anxiety might regularly cast an eye around the camp and fret about all the accidents waiting to happen. Here, we have a teenager sitting right by the trail, making arrowheads. *Do that at the flint-knapping site so people don't slice their feet open on your flakes.* And there, we have a basket of meat sitting in the sun. *We'll all get sick if that's not cured over the fire.* Here is an empty water gourd, and there are crocodiles in the river. *Get water before dusk and, on the way back, more sticks for the fire, just in case . . .*

Orderliness is part of a long-term strategy. By anticipating problems and preventing them, an orderly person clears the path ahead. The effort she invests today may not have an immediate benefit. In fact, as she carves a mark into the trees she passes, she loses time in which she could be gathering food. But when the sun begins to drop, she'll save time by following her own well-blazed trail home. And tomorrow she'll save time on both ends of her commute.

The orderly personality foresees problems, and leaves less to chance. Of course, that arrowhead cuts both ways. If you stick to your beaten, blazed trail, you may miss out on wonders and astonishments found only on the road less traveled.

And it's not uncommon for an orderly personality to smoke the meat, pile up the firewood, fetch the water, and clear the yard of arrow flakes, only to see the river rise and wipe out the entire campsite. Time to move on. The best-laid plans of mice and men oft go askew.

And so, as with anxiety, it's nice to have a middling amount of orderliness. Too much, and you exhaust yourself preparing for troubles that never arrive. Too little, and your spouse bleeds out after stepping on arrowhead flakes, your whole family gets dysentery from bad meat, the fire dies from lack of fuel, and as you all stagger about in the dark to relieve your spasmodic bowels, you're eaten by wolves.

Prior to psychiatry and effective drugs, evolution must have weeded out the extremes of personality. The personality that is wildly disorganized, along with the personality that cannot cease washing her hands at the river, will not fare well in the competition to raise many healthy children. And dead children is how evolution has traditionally set limits on how extreme any personality is allowed to be. It's nature's way of keeping good order.

So You Think You Might Be Conscientious

People make money in the stock market two ways. People lose money in the same two ways. The two ways are long-term investing and short-term investing.

I subscribe to the classic long-term strategy. Years ago I bought a bunch of stocks. I didn't choose randomly. I wanted to be a part-owner of companies that behave as conscientiously as I do. After much research I invested in a half-dozen corporations with good reputations for respecting their employees and the environment.

That was when the market was high. And it went even higher. But then it went low. Really low. On paper my portfolio sank to a fraction of its former value. Like a sullen fish it hit bottom and hunched in the muck, scarcely fanning its gills. For a few years my portfolio hunched there as stocks leaped and flopped and died around it. And then grudgingly a few of my own holdings stirred. My portfolio inched forward. Today it's no aquarium prize, no example of vibrant, robust investing. But it's alive.

I never considered selling. That would have been irresponsible. I had chosen carefully, and I was committed to my choices. To sell my stocks would have meant accepting a loss, and that simply was not an option for my personality. Furthermore, it would have required further research into replacement stocks, and I knew I didn't have time to perform that task properly. If I hastily traded the devils I did know for devils I didn't, I could end up with a far stinkier fish. So I bought and I held. And held. And am holding still. Someday that portfolio is really going to amount to something. Just you wait and see.

A sugar rat I know subscribes to the short-term approach. If he weren't gainfully employed, he would make a smashing stock trader. He's wired to notice the smallest change in his environment. When a fish begins to swim, he pays attention. He reads about that fish and monitors its movements with the intensity of a cat watching birds through the window. Only when that fish settles again does he relax. His attention wanders. And then a new fish catches his eye.

If he could devote substantial time to this, he'd probably make a fortune. Or not. Some traders do make millions by consistently swapping sluggish fish for others that seem poised for a spurt of activity. But the more common pattern is to bet on a fish who turns out to be merely scratching his fin, not readying for a sprint.

It's easy to lose, either way. Neither strategy offers an obvious advantage. And that's why both personality types walk among us today.

I'm thinking of my aunt and uncle's chipmunk, Microchip. He lives somewhere under their barn, and transfers bushels of sunflower seeds from Janet's hand to his storerooms. He has a long-term strategy. All winter he'll be spared the danger and discomfort of foraging for seeds. Without even taking off his slippers he'll be able to shuffle from bedroom to pantry when he's peckish. That's planning ahead. That's the dutiful, cautious thing to do.

But there is also a rat, a regular thug of a rat, who barges through a loophole in Microchip's long-term strategy. The rat is not a big fan of food storage. The rat seizes the day, seizes the territory, and seizes Microchip's seeds. That's making hay while the sun shines. That's the spontaneous, opportunistic thing to do.

Once again, there's no obvious winning strategy. The chipmunk risks months' worth of effort. But the rat risks going hungry every day.

I'm not suggesting that a personality low in Conscientiousness is a rat in the ratty sense of rats. The average person with low Conscientiousness does not take advantage of the people around her, or exploit their hard work for her own benefit. She just follows her bliss a little more assiduously than others might. Her brain is oriented more toward enjoyment than obligation.

On the mild side of low Conscientiousness, a person will neglect to keep pace with her own goals. She might not do her homework, or her laundry, or pay her bills when she had intended to. *I should, but let's go see a movie.* As Conscientiousness dips lower, it starts to affect others. The dog goes unwalked, the bake sale doesn't get organized, the coffee break lasts forty-five minutes. *I know, but I just don't feel like it.* Things get really unpleasant when Conscientiousness reaches a level that permits a person to intentionally cheat others. Now the bake sale doesn't get organized—and it's someone else's fault. The homework doesn't get done—but the student insists she passed it in. The coffee break gets longer, even after a warning from the boss.

Heading the other way on this spectrum of personality, moderate Conscientiousness seems like a nice place to land. One would behave reliably enough to enjoy the respect of her community, but would be able to slack off occasionally without feeling like an utter parasite.

As Conscientiousness rises higher, flexibility falls. The very Conscientious personality tends to see only one way to do things: The right way. This can narrow one's own thinking. And it can be really annoying to others.

Let me just state for the record, as a card-carrying drudge, that we the

Conscientious do notice when you cut corners. We see you leave the milk out. We know you didn't do your homework. And because we're so Conscientious we feel obligated to report your supersized coffee breaks to the authorities. If the low Conscientiousness personality goes by the nickname "Slacker," the high Conscientiousness personality is known as "Nitpicking Pain."

But high Conscientiousness delivers some nice features, too. Conscientious people reap a health benefit from all our obligatory toiling in the areas of diet and exercise and brushing and flossing. Conscientious people also tend to rank high in "morningness," adding to our reputation for wholesomeness and being better than thou. Even the spouse of a Conscientious person lives longer than average, presumably thanks to all the beneficial nagging. (You're welcome!) And Conscientious people report an extra measure of "commitment" and "intimacy" (trust) in their marriages. (OK, and also the tiniest deficit in "passion.")

The biggest drawback attached to this personality factor is the potential for addiction that comes with low Conscientiousness. And it's a doozy.

As Marilyn Carroll's rats demonstrate, a dopamine setting that's stuck on *MORE* is susceptible to hijacking. Anything that can trip the system's trigger has a shot at becoming an obsession. And unfortunately the modern human environment is rife with candidates. Dopamine-tickling sugars and fats, alcohol, cocaine, heroin, meth, nicotine, ADHD meds, poker games, video games, and countless other enticements beckon all along the path that a modern human treads on his daily rounds.

This liability is among the worst for any personality facet. I grew up in a home where addiction dictated the tone of all our interactions. It wasn't great. And many of my friends are in recovery from one addiction or another. I've heard their stories. But the full horror of runaway dopamine struck me when I found an Addiction Severity Index. *Cool! I wonder what my score is!*

But it wasn't cool. It was enormously sad. If you have a dopamine system that knows how to say "That's enough," this test will make you feel like the luckiest person alive. Addiction is so unfair. And so fierce. Addiction is a progressive brain disorder like Alzheimer's. The addict's life, without treatment, spirals down into mental and physical chaos that also afflicts the people around her. And so the questions on the Addiction Severity Index are anything but amusing.

How many times in your life have you been treated for drug abuse? (The failure rate is only moderately more successful than the treatment for Alzheimer's. And so it's not "have you," but "how many times.")

How many days were you paid for working in the last 30?

How many times have you been hospitalized?

How many times have you been arrested?

Do you have a close relationship with the following people: Spouse? Children? Friends?

How many days in the past 30 have you had serious conflicts with other people?

And on, and on. The questions are all grim. They're all about loss and pain and conflict and sadness. None of them are, like, Where would you go on vacation if money were no object? Who's your favorite football team? Do you dislike Brussels sprouts but eat them anyway to be polite?

After taking that quiz I vowed, for all the addicts in the world, that I would try to complain about nothing for the next twenty-four hours. By chance, I am blessed with a personality that resists the addictive brain baubles surrounding us in the twenty-first century. I'm tremendously grateful for that.

Happily, nobody's personality is written in stone. Science shows us that about half of the variation in any given trait is genetic. That leaves the other half susceptible to the forces of our environment. And even if the environment of childhood brought out your worst, it's not too late. While Carroll and other scientists search for better ways to pry an addicted brain free from its fixation, the rest of us can employ old-fashioned character development.

For the high-Conscientiousness harpy like me, it is useful to love other people as they are. This is not easy, since we tend to do things right, and that means everyone unlike us is doing things wro— Ahem. What I mean is, I'm learning to recognize the sensation of my brain going rigid as it locks on to a vision of how things should be. That feeling is a reminder to pry open my mind to other possibilities. After all, human personality is diverse today because that has successfully served our species for gazillions of years. Who am I to say one form of our diversity is preferable to another? The very existence of the diversity would prove me wrong.

Furthermore, if every personality were as dutiful and cautious as mine, who would have invented fireworks, frosted peanut-butter-cup brownies, or the unicycle? And even if someone as logical and reasonable as I somehow managed to invent those impractical items, my life would be insufferably dull without my spontaneous friends. Without the sparkle and drama of the people around me, I'd bore myself comatose in a matter of days. And on that one day a year when I threw caution to the wind and left my desk to have lunch in town, whom would I call? All my friends would be chained to their own desks. I strive to appreciate human diversity, and I am getting better at it.

On the low-Conscientiousness end of the spectrum, change is possible, too. A mountain—nay, a mountain range—of self-help books with titles such as *The Now Habit, Getting Things Done,* and *Order from Chaos* implies that a whole lot of people wish their brains were a little more Conscientious. The same mountain range suggests that nobody has perfected a method for cultivating the characteristic.

Because of the addiction risk, researchers aren't giving up. Taking no chances with her own children, Marilyn Carroll deliberately tossed temptations at their dopamine systems to see what might stick. If their personalities turned out to want *MORE,* she was determined to get them hooked on something healthy.

"You can compete with the addiction with things like social events, exercise, sports," she says. "We kept our kids in every sport they could be in. They were in band, camp, they were constantly doing things."

One recent study found that people with disobedient dopamine can benefit from a strategy of anticipating weak moments. Rather than staking one's hopes on never slipping, the researchers proposed a system of planning for the inevitable: *If I skip the gym tomorrow, I'll go twice on the weekend. If I have a drink, I will call my AA sponsor. If I eat six frosted peanut-butter-cup brownies, I won't eat seven.*

The most reliable antidote for now, however, is time. Like impulsiveness, which burns brightly in adolescence then mellows with maturity, Conscientiousness also ages well. The average personality becomes more industrious and reliable with age. Perhaps this is the influence of environment, of cultures that prize hard work and honesty. But it may also reflect an aging process inside the brain, as waning hormone levels alter the activity of dopamine and other neurotransmitters. In any case, turning a blind eye to temptation gets easier with every passing year.

5

FACTOR:
OPENNESS

OPENNESS FACETS
Intellect
Imagination
Artistic Interests
Liberalism
Emotionality

THE CAST

Mascots ....Doogie, Harris, and Shackleton

Neurotransmitter ....Dopamine and a cast
of thousands

Brain Regions ....Hippocampus and
Left-Lateral Temporal region

OPENNESS

HIGH OPENNESS PRODUCES A flexible personality. Such a person welcomes new ideas, even if they clash with older beliefs. If this means changing her view, that's fine. This person's brain thrives on new sights and sounds, and enjoys all kinds of mental exploration. A person with low Openness is not closed-minded, she simply doesn't crave so much mental stimulation. Her mental strategy favors stability over flexibility. She's satisfied with the concrete world, and isn't prone to flights of imagination or late nights discussing the meaning of life.

In a way, Openness looks like Extraversion. But Extraversion is a pattern of throwing open the doors and walking out through them. Openness is a tendency to throw open the doors and invite the whole wide world to come in. The "approach" energy is what they share. High Openness indicates an embrace of mental stimulation and mental exercise. An Open personality is attracted to ideas, the more unfamiliar, the better.

This factor also has elements of the "avoid" orientation that you'd find in a Neurotic personality. A low-Openness personality is put off by uncertainty, and finds security in protective rules and regulations.

The facets under Openness run to the sophisticated side. A mouse may not display the finer points of Openness. That said, they certainly have intelligence. Some mice are born with greater powers of learning and memory.

I believe mice have artistic interests, too. Their brains and bodies both thrive in an environment that offers objects to look at and manipulate. And like any museum patron, they like the exhibit to change regularly. I can even make a case for magical thinking and religion in mice.

That shouldn't be too shocking. We all, mouse and man alike, evolved personality in response to the challenges thrown at us by our environments. But we, with our complex brain, do go a bit wild with this factor. Despite wide variation among individuals, as a species our intellect is stunning. Our imagination likewise is worlds away from that of other creatures, as far as we can tell.

Emotionality and liberalism are particularly relevant to how we get along in today's environment. Some of us can effortlessly read emotions in others, and respond in a way that builds a cooperative, helpful relationship. Others are tone-deaf in this language, leaving them to focus on their own business.

The facet most primed for conflict is liberalism. Most industrial cultures now govern themselves with laws and elected lawmakers. This pits two extremes against each other: Liberal personalities that see every human as equal clash with low-liberalism people, who have greater faith in laws than in people.

Openness Facet:

INTELLECT

INTELLECT INDICATORS	RARELY	SOMETIMES	OFTEN
I LIKE RIDDLES AND PUZZLES	❏	❏	❏
I DON'T MIND FIGURING THINGS OUT FOR MYSELF	❏	❏	❏
I CAN IMPROVE ON OLD WAYS OF DOING THINGS	❏	❏	❏

THIS GIVES YOU A QUICK look at where you land on this facet. If your answers tend toward the "often" side, you're higher in that facet.

These questions elicit your intellectual style. People who score high have restless brains that seek stimulation, or as one intellectual scientist has phrased it, they "enjoy effortful cognition." Just as some people feel great when they exercise their bodies, these people are exhilarated by mental exercise. People with a low score find more gratification in real people and things than in abstract ideas. Although this facet of personality doesn't claim to measure your IQ, there is a small overlap: Statistically, people who love to exercise the mind also score a little higher on IQ tests.

Intellectual Mouse

The Five Factor Model of personality measures your intellectual metabolism, not your IQ. But since mice don't demonstrate much in the way of intellectual style, let's look at general intelligence anyway. There we'll find enough brilliant mice to stock an Ivy League university, along with some of the most haunting personality disorders known to man.

John Roder knows a thing or two about both subjects. He's the landlord to a new member of Mouse Mensa, a creature who can solve a maze in half the customary time. And ironically, Roder's own personality is beginning to fray at the edges, a bitter affliction for a lifelong scholar of the human brain.

I find him high in a tower at Mount Sinai Hospital in Toronto on a foul November day. After a pell-mell morning of air travel, I'm cheered to discover a vending machine between the elevator and his office. But as I approach to pick my poison something seems off. The keypad looks like the U.S. variety, and a blinking slot beckons to my credit card. Various products wait in their spiral pushers. But those products are…entirely unfamiliar. The fine print on the boxes is disappointing: Shrimp alkaline. Calf intestine. T4 DNA Ligase. The vendors of these lab staples have found that self-serve machines work better than salespeople. If the customer needs calf intestine, he steps into the hall, swipes his credit card, and gets calf intestine. But if the customer needs a soft drink, I guess he goes to the cafeteria.

When I enter his office, Roder twirls off his chair and lurches forward to shake my hand. His eyes rove over me and I'm not sure he's smiling. There's something tight about his face, his expression. His words make sense but they come out in slurry bunches. Oh, well. Onward.

He has lined up a full afternoon of meetings with his lab researchers, and my airplane was late. So he throws together a cup of tea and herds me back out the door. His feet flap and his gait is stiff. Tea droplets trail us down the hall. I don't want to stare, but I am a science writer and there's some science going on right beside me.

Soon enough I'm distracted by Steve Duffy, whom Roder has asked to teach me about learning: how the cells and chemicals of the brain store information. It's a question I've marveled over many times in my life. My memory holds so many things, so many facts, so many voices, places, smells, faces, the multiplication tables, the lyrics of a thousand songs, that jar lids turn left to loosen, a smattering of birthdays, and the subway lines that go to Brooklyn. How, how, *how* can it all fit in my little brain? And in such detail? I could walk you through the house I grew up in, room by room, describing the color of the walls, the texture of the furniture, the pattern on the carpets. I could give you a decent tour of my first-grade classroom, including the smell of the chalkboards and the taste of my peanut-butter-and-honey sandwich.

How? How do we learn this stuff, and more baffling, how do we remember it?

"It's a pretty simple phenomenon," says Duffy, a solid guy wearing bold, black glasses. Oh, I bet it's simple.

"Neurons excite," he explains, "mainly using glutamate, causing the cell to depolarize. That's a drop from seventy millivolts to sixty millivolts." He keeps going, but I don't even write down the words.

What he's describing is "long-term potentiation." This goes back to the synapse, the intersection between two (or among three or four) neurons. For your brain to function, information in the form of chemicals and electricity must cross that intersection from one neuron to the next. Normally, when a signal needs to cross, the sending cell tosses chemical keys into the intersection. Some of those keys open the locks on the receiving cell, and the signal is allowed to continue on its way. Then the gate closes until it's needed again.

Long-term potentiation is a formal way of saying: When you learn something, it's like the gate doesn't close. That intersection is now dedicated to a particular bit of information. *Yoga every Tuesday. Don't combine bleach and ammonia. Orange is the new black.* That's the learning part. When a signal crosses the synapse intersection a few times, it changes the synapse.

If you're thinking that there simply aren't enough nerve cells in the brain to dedicate one to each minuscule detail, here is relief for your fretting neurons: Each of your 100,000,000,000 neurons can have 7,000 synapse points. These nerve cells aren't smooth and straight like tulip stems, with a blossom at one end and a bulb at the other. They're more like thorn trees. The thorns are called "spines," actually, and each spine has the ability to play the synapse game, if needed. This opens a lot more real estate for your knowledge to settle into.

And that's just the physical part. Learning also depends on a brainful of chemicals such as NMDA, glutamate, ERK, CaMKII, PMK and some things whose names incorporate symbols not found on a modern keyboard. There is nothing simple about learning.

Perhaps this is why there are so many smart mice running around. Each animal embodies a different detail of the learning process.

Doogie was the first. A winsome brown creature, this mouse was "published," as geneticists say, in 1999. In Doogie's head, scientists had found a way to inject a chemical doorstop. This was long-term potentiation on steroids. Long after a normal mouse's synapse gates would have eased shut, Doogie's gates would remain jammed open. With the path thus cleared for electrical impulses, Doogie was a champ at learning mazes. "Spatial learning," that was his strong suit. He solved the Morris water maze, where a mouse swims in a pool until it locates a platform just beneath the surface, in three tries, as opposed to a normal mouse's six. Doogie blazed through mazes.

More recently, in 2005, a team at the University of California in Los Angeles accidentally created a mouse I'll call Harris, after his altered "Hras" protein. The researchers were working on a rare human nerve disorder that clogs nerve cells and slows learning. By chance, a protein molecule they were tinkering with in mice altered the speed of a mouse's nerve firings. In a mouse with his DNA altered to exaggerate this protein, neurons fire like machine guns. In the synapse, an avalanche of keys tumbles over the locks. The gates fly open and the learning goes through the roof.

Harris demonstrated his smarts on the foot-shock apparatus. I think of this as the "hot asphalt protocol." I was pretty young when I first set a bare foot on hot asphalt. Wow! Ow! I hopped, I squeaked, I hot-footed back to the grassy road shoulder. And ever since, my brain has associated hot tar with stinging feet. I steel myself for it, and then dash across, squeaking and hopping. The foot-shock test measures how quickly a mouse learns the same lesson.

I saw the test in action in the Lesch lab in Germany. A technician whisked a black mouse from her condo to a wire test chamber surrounded by calming black curtains. The researcher, Thomas Wultsch, gave her a few minutes to sniff and whisker her new environment. Mice don't learn well if they don't have time to settle down first. Eventually Wultsch cued a tone that played in her chamber. *Hoooot.* There was a pause. And then the mouse was hopping around the perimeter of the chamber the way I hop on hot asphalt. The bars of the floor had been mildly electrified. "Squeak! Squeak!" she protested, much as I squeak when I'm hopping on the asphalt. The shock ended. And she returned to her condo and her comrades.

Sometime later, we airlifted her back to the chamber. Once again she whiskered around in a mousy manner: sniff, sniff, poke, poke. And once again Wultsch sounded the tone. *Hoooot.* And she froze, an instinctive response to danger. She had learned, after just one experience, that the tone was paired with a shock.

But Harris learned even faster, and from even more subtle lessons. This mouse was given no time to explore the chamber before the tone and the shock, and his shock was so mild that some mice wouldn't even bother to remember it. He shouldn't have remembered the event at all. But he froze immediately the next time he visited the chamber, not needing even the tone to jog his memory. Harris also trounced normal mice in the Morris water maze.

Doogie and Harris carry completely different mutations, but the mutations improve the same brain region, the hippocampus.

"It's a tiny region," Duffy says, showing me a diagram. In a mouse, it's a banana-shaped structure occupying perhaps 10 percent of the brain. "But it's the center of learning and memory. It's also prone to stroke and epilepsy, which is why those events often leave you without short-term memory."

Is he telling me that all my song lyrics and names and faces and smells and the pattern on my childhood carpets are stored in that little thing?

No. Phew.

"We think the hippocampus is like a phone book," he explains. "It sends all that data to the cortex [the outermost brain layer] for storage. When you need to remember something, the hippocampus calls it up." Complex memories, like my childhood kitchen, may be broken up into components: The spatial location of the Atlantic cook stove is stored in one area of my cortex, while the appearance of its glossy gray enamel is shelved in another, and the velvety texture of the basset hound dozing beneath it is on yet another shelf. Duffy thinks of such memories as a "constellation of neurons," all linked by synapses and twinkling to life as the hippocampus calls on them.

So Doogie and Harris, and presumably a brilliant human being, all have a high-performance hippocampus. It's small, but mighty. Without that librarian directing the flow and storage of information, we'd all be lost.

Conversely, if we could learn how to improve the hippocampus, we might be able to help damaged brains to learn better.

While Duffy was filling me in on some of those interactions, Roder had flapped out of the room. He returned with a Greek god. Again I abandoned my note-taking, this time to indulge in some staring. This youth's features approached a perfect symmetry. His mouth was a cupid's bow; his dark curls were tied in a pile on his head. If symmetry does reliably indicate a slight increase in intelligence—and science does suggest that is the case—this young man may go on to cure Alzheimer's, schizophrenia, and middle-aged forgetfulness. For now, Roder was guiding him toward a PhD.

His name is Bechara Saab. And he and some colleagues have gathered up a few promising scraps of research that were lying around the lab, and kneaded them together to see what might emerge. What emerged was a mouse with a great deal of curiosity. And that curiosity emerged hand-in-hand with speedy learning, and excellent memory.

Is curiosity the key to intelligence? This has some appeal. Curiosity leads animals forward to discover new possibilities. In simple animals, those possibilities will be tangible, physical things like food and friends. But in humans, curiosity could also inspire the intellectual journeys we call "effortful cognition," imagination, and artistic interest.

So what does curiosity look like in a mouse? It's a hippocampus that responds to the lightest touch. It sings like a Stradivarius. It accelerates like a Ferrari.

Dopamine plays a big role in the sensitivity of this refined hippocampus. I find that noteworthy, given dopamine's sponsorship of Extraversion—especially Extraversion's experience-hungry, impulsive characteristics. Furthermore, we just saw in the Conscientious chapter that dopamine produces the drive to consume the things we require to live long and procreate. Without dopamine, a rat won't seek food, or even pick up food that's under her nose, although she eagerly chews food if a scientist puts it into her mouth. Now with dopamine also starring in curiosity, it's starting to feel like the "go for it" chemical.

In Saab's mouse, dopamine receptors—the locks, not the keys—appear to cause curiosity. And it's only the dopamine receptors in the hippocampus, that library of learning, that are different. What Saab did to create this mouse was add more dopamine locks specifically to the hippocampus. Rather than slowing the flow of information, he anticipated this would increase the number of gates available to an incoming signal.

And how did this impact the mice? They proved to be little Marco Polos, little Shackletons. When Saab placed these mice in an unfamiliar box with unfamiliar objects, they spent a lot more time whiskering around than normal mice. And they did a huge amount of rearing. In a scary environment, mice rear out of fear. But Saab was offering them a secure, dimly lit box. And when mice rear in a safe environment, they're exploring. These mice reared twice as much as others. Their appetite for information was enormous.

Saab gave the Shackleton mice another test of curiosity. In a maze with many arms to explore, the curious mice visited twice as many arms as normal mice. And instead of taking five minutes to acclimatize before exploring at all, these mice set forth after just two minutes.

These mice are not hyperactive or impulsive, traits that can also flow from dopamine. They're not running around because they're restless. Saab ruled out that possibility by testing them in bright light, which is scary to a mouse. Like normal mice, they declined to go poking around when their instinct told them it was unsafe.

Nor were the mice exploring mindlessly. They were learning about the places they went. And they were doing it with great speed. Saab gave them and some normal mice one trial in the Morris water maze, then returned all the mice to their condos for twenty-four hours. When he gave them another try the next day, the normal mice had all forgotten where to find the platform. But the Shackleton mice remembered enough that they spent most of their time searching in the correct quarter of the pool.

What was happening in those brilliant little brains? To find out, Saab sacrificed some mice and studied fine slices of their brains.

Brain-slicing I had also seen at the Lesch laboratory. A technician, hunched over a cooler, worked with a white brain the size of a peanut. Deep-frozen so it would hold its shape, the brain slid past a cold blade that shaved off a slice. The technician blotted this tissue, crumpled like potato chip, off the blade with a glass slide. He stuck his warm human fingertip to the bottom of the glass. As the slice thawed, it relaxed onto the slide. An entire brain would be sliced thus, creating a front-to-back encyclopedia.

Selecting slices from the hippocampus of the Shackleton mice, Saab's team made a nerve-by-nerve study of how this curious brain was learning. With tiny instruments, they would try to mimic learning by shooting tiny electrical currents across a synapse. This would tell them if the Shackletons' neurons were any more sensitive than normal. So they set the current for half the strength normally required to bridge the gap, and they hit the switch. In a normal mouse this would be able to open only enough gates to create a "short-term potentiation," the faintest trace of a memory. In the curious brain it nearly blew the gates off their hinges. These brain slices were recording indelible memories right there on the bench.

And so another piece of the puzzle falls into place. There will be many more. And the discovery of the pieces won't guarantee we'll be able to use them to improve human health. I mean, it's great to know that adding dopamine receptors to your hippocampus would probably make you a curious genius. But how would we do that, exactly? In the Shackleton mice, it entailed gene manipulation. For now, humans aren't lining up to get our genes rearranged.

And we don't know if the Shackleton mice are handicapped in some other area. Roder is back, spilling tea and listening with intense interest to his student.

"There are different types of intelligence in mice," he interjects. "You can have a strain that's high in learning and memory, but low in sociality, for instance. I don't have social intelligence," he adds with a grin. "I appreciate it in others, like my wife."

Intelligence, in mouse or man, is much greater than the sum of our recorded bits of data. Intelligence rests heavily on our ability to rearrange those bits of data to create new ideas. This is what allows animals to navigate an environment that's always changing. This is general intelligence, or g.

When I brought Mitzi and Maxi Mouse home from the pet store and put them in their new condo, they fell back on g. They had never seen a water bottle exactly like the one I got them, but they went straight to it and drank. Nor had they seen a wheel exactly like mine, but both mice climbed aboard and began to run. (Mitzi, my anxious mouse, didn't stop for about twenty-four hours.) At the pet shop they had slept in plastic igloos, but here they figured out how to burrow into a pile of shredded bank statements. The environment was unfamiliar, but calling on their memory of previous environments, they generalized.

General intelligence allows us to adapt. The first time I walk into your house I, like my mice, will know how to get water. Your tap may look different from any I've seen, or you may have just a barrel sitting in the corner. You may have cups, glasses, or a ladle. Regardless, I will be able to rally old memories and put them together in a way that allows me to master this new situation. It's an efficient system because it allows animals to apply general rules to their environment, instead of memorizing each detail.

Scientists can make a mouse who learns well but then forgets, losing his *g*. They can make one who solves a maze quickly but then can't unlearn that route when scientists move the reward to a different spot in the maze. Or they can make a mouse who learns well but can't use her knowledge because she's too hyperactive to concentrate.

"There are dozens of smart mice," Saab had said. "What's great about it is the specificity. We know exactly what we're changing. It's getting to the bottom of how the brain works, how information moves around."

But some mice are better learners than others even before scientists fiddle with their genes. Why hasn't evolution made every mouse a little genius?

We brainy humans forget that smart is as smart does. An animal needs only as much brainpower as it takes to successfully raise offspring. For most animals, life's challenges can be surmounted with a minimum of wattage. And extra wattage may carry a risk. A smart mouse will learn quickly about her environment, which allows her to do her chores more efficiently. But learning demands attention. For her entire life a smart mouse will devote more of her attention to updating her memory banks, and adding new landmarks to her mental map, and noticing new shortcuts to her feeding grounds.

Attention is not free. A brain can only do so many things at once. And a mouse who is jotting down landmarks and collecting odors is not paying as much attention to predators as a simple mouse might. The simple mouse learns one route to work and follows that same route until the day he dies. Because he's on autopilot, he can devote his brain's full attention to hawks, owls, foxes, snakes, and weasels. So why hasn't evolution made every mouse simple? The price a simple mouse pays comes due when his environment changes—someone wipes out his seed patch, and he has to find a new one. And then remember where it is.

Each lifestyle works adequately. Neither works flawlessly. But because a diversity of personality styles gives the entire species a better chance of surviving nature's challenges, that diversity persists.

Intellectual Human

John Roder was not always so interested in intelligence. He began life studying immunology before veering into neurology. For a while his interest was in the nuts and bolts of learning, identifying the molecules and neuron types involved. And then his son developed schizophrenia. It arrived right on schedule, as the boy crossed the hormonal bridge to adulthood.

"Schizophrenia has two parts," Roder says in his rushing style. "The hallucinations can be treated. But the other part involves memory and attention. None of that can be treated. He's kind of frozen in time when he got sick."

Schizophrenia reshapes the personality of one in a hundred people. Almost all of us know of the disease through a friend or family member. And for the part of schizophrenia that disrupts learning, there is no drug. Even the drugs that quell the psychotic visions and delusions are so burdened with side effects that many people try to live without them. We desperately need new drugs. And that means we need schizophrenic mice.

Scientists have identified more than a hundred genes that probably contribute to schizophrenia. Roder, with a new sense of urgency, went after the one that had already been confirmed in humans. The gene, named "Disrupted in Schizophrenia 1," turned up when researchers sifted the genes of a Scottish family in which the disease ran rampant. Roder wanted to know what those disruptions might be, and whether they affect a mouse's learning and memory.

With a molecule-sized blunderbuss, Roder took aim at the gene in a Black 6 mouse, and blasted it to smithereens. When nature isn't producing mutations fast enough, an impatient person can hurry things along. And then he can paw through a gene's wreckage to see if any of the induced mutations look interesting.

He bred 1,686 of these mice, and screened the DNA of every one, looking for mutations that might warrant further research. In that haystack of data he found two mice he wanted to know more about. He bred a herd of each. As he put the two strains through their paces, one of them began to look like his son.

Like a human with schizophrenia, Mouse 100P was hard to startle. His brain seemed so overwhelmed with information that he hardly noticed noises that startle normal mice. The mouse also appeared to have the too-open attention that is responsible for hallucinations in humans with schizophrenia. And when Roder issued 100P the same drugs that help schizophrenics, the mouse's attention improved. Roder was on the right track. But we already have mediocre drugs to control hallucinations. Roder's main interest was the symptoms for which we don't even have mediocre drugs.

He turned his attention to 100P's memory. In a simple "T" maze, mice had the option to turn left or right to find food. Could the 100Ps remember which way to turn the next time they were in the maze? Not on the first day. And although they did get better on the second day, they were only half as good as normal mice. Day after day, as normal mice got better and better at remembering which way to turn, the 100P mice fumbled through the decision. After ten days their performance hit a plateau at 70 percent correct—a point the other mice had passed a week earlier. Mouse 100P can learn, but it takes him many more tries than a normal mouse. Etching a memory in this mouse is like digging a trench with a popsicle stick. It must evoke a mixture of excitement and sorrow to create a new mouse that is useful because its disordered behavior matches your son's.

And now the question is, what makes 100P disordered? What, among the synapses and the thousands of different chemicals, is slowing the flow of information? Why isn't information being properly shelved in the memory?

Learning and memory are so complex that the 100P mouse may end up helping schizophrenics, but no one else, to improve their thinking. The same goes for Alzheimer's mice, of which there are multitudes. That disease has its own specific glitches, and deciphering them probably isn't going to shed much light on schizophrenia. But gradually, as scientists around the world break apart the brain, we're learning about the tiny interactions that add up to human thought.

My culture reveres intelligence. We tend to think it's always a blessing. And it often is. The environment that we have created in the industrialized world—an environment popping with machinery and governments and concepts such as an infinite universe—demands a lot of firepower. It takes a heck of a lot of g to get through a day in the developed world.

But after a point, intelligence is as intelligence does. High intelligence does not guarantee you a trip to the top of the heap. I recently read an editorial in a science journal complaining about this, lamenting the rise of the "dull scientist." The academic system, the author argued, rewards Agreeableness and Conscientiousness at the expense of recruiting intelligent, creative people who are a pain in the butt. The *g* doesn't stand for great.

Everybody knows a personality like that—a personality who can think circles around you, but you don't really want to be around. My father was that sort, actually. One of the most curious and analytical people I've ever known, he sometimes exhibited a level of social intelligence you'd expect to see in a badger.

Beyond a certain level of intelligence, it's not how much we learn, but what we learn that makes us who we are. And what we learn is determined by all the facets of our personality. Perhaps your personality causes you to be interested in skydiving and foreign films, and those things shape your memories. Or your personality points you toward beach time with your buddies, and coaching little league, and your memories of these experiences add weight to your personality. Your personality determines what you learn.

If you could stick your finger about two inches into your ear, you'd touch your hippocampus. It lies like a lumpy jelly roll near the bottom of its hemisphere. At the front end it bumps against that ancient emotion center, the amygdala.

The hippocampus does much of its work unaided. It accepts information, records it, and shelves it out in the cortex. But every so often an animal learns something that it should really, really, definitely remember. This isn't multiplication-table stuff. This is you bitten by a dog. This is your mother's shriek as you run into the path of a truck. This is the taste of ice cream.

When the amygdala registers such potent events, it takes immediate action: *Scream! Stop! Eat!*

But it also tries to improve your long-term prospects. It stamps URGENT on the experience, and forwards it to the hippocampus. As Steve Duffy put it, the hippocampus is "dry." The hippocampus doesn't judge. It dispassionately manages information. The amygdala is soaking wet. The amygdala is about urgent emotions. And anything the amygdala deems urgent gets special treatment in the hippocampus. It's this sort of interaction that personalizes our learning.

I'm thinking of my own amygdala, the almond-shaped alarmist between my temples. It goes ballistic for every popped balloon, every angry face, every wailing baby. The sensitivity of my amygdala means that my hippocampus receives a lot of data stamped URGENT. And the hippocampus doesn't judge. It just files. The shelves of my cortex are loaded with memories of the scary and the sad.

A typical conversation between my sister and me:

Me: Remember Dad betting us that he could eat a whole spoonful of Tabasco sauce, and Chip sneaking downstairs to shut off the water?

Sister: Noooo...Remember when he said there was a witch in the kitchen and then smoke and a bang, and silver dollars went flying everywhere?

Me: Noooo...I have a silver dollar, but...no memory.

The Tabasco episode is seared into my memory because two aspects of the event alarmed my amygdala. One, Dad's temper was unpredictable, which scared me. And two, his face bore an expression of agony as he stumbled toward the tap (which yielded only a gurgling sound). Memorable.

You would think I'd recall the explosion, but the alarm must have been calmed quickly by laughter. No one wore a pained or angry expression. Bright silver dollars were bouncing off the walls. So, who cares? What's to remember?

It's the same reason I never remember where I parked my car when I come out of the grocery store. Nothing could be less emotionally stimulating. And so I wander like an amnesiac, craning, squinting. On the other hand, I'll never forget the three times I've lost control of a car. Those events were plenty stimulating. The color of the car, the glassy sheen of the ice, the cars around me, all were noted by my amygdala and forwarded *urgently* to the hippocampus. They rest on a shelf marked NEAR DEATH EXPERIENCES, along with the minutes when my cousin slipped and fell into the ocean in the winter and couldn't get out, and the time my humorless grandfather spanked me, and a certain airplane flight in Madagascar...

And so that old approach/avoid dimension of personality influences what you notice, and what you remember. Glimmers of this emerge from various studies: Neurotic and shy people are quick to memorize an object if they see someone else looking at the object with a scared expression. But they don't remember objects that were looked upon by a happy person. By contrast, people who are low in Extraversion but not wildly Neurotic do seem to get a memory boost from someone else's expression of happiness. Anxious people are particularly sensitive to subtle cues around them that indicate trouble: They're quick to spot an angry face in a sea of neutral faces. Extraverts, on the other hand, have amygdala that shrug off the appearance of a strange face. Perhaps my husband's inability to remember names arises from the fact that meeting strangers causes him so little anxiety. The data—Jack, John, Jim—passes his amygdala unchallenged, and enters the hippocampus with no rating of importance.

You can start to see how your personality shapes your memories. And this stored data adds up to the personality you present to the world. The way you greet a friend is not instinctive. It's learned, remembered, and called up according to your personality. It may take the form of an Extraverted whoop, a Neurotic whine, an Agreeable compliment, or a Conscientious show of punctuality. Your greeting displays your personality.

Circling back to the personality facet of intellectual style, that, too, influences the diet of information that you feed to your hippocampus. Humans, with our tremendously complex brains and personalities, vary a great deal in how much we enjoy giving our hippocampus a workout.

If your brain craves exercise and stimulation, your hippocampus will work like a dog. It will spend its jelly-roll life shuffling information from hither to yon, sorting and filing, sorting and filing. As you put your feet up for an after-dinner discussion of the infinite universe, your hippocampus will scurry and sweat, snatching up one stored notion after another to keep up with your flapping jaws. And you'll feel great.

On the other hand, if your brain prefers predictable foods to mysterious ones, and you'd rather read a nice novel than take a class in Ayurvedic massage, then your hippocampus will work at a more leisurely pace. When you shoot hoops or take a drive after dinner, your hippocampus will put up its feet and relax. And you'll feel great.

Evolution of Intellect

Why intelligence evolved is pretty obvious: It helps an animal solve problems, and for most animals, life is one long chain of problems. More smarts begets more success begets more babies.

More puzzling by far is: Why low intelligence? Why isn't every snail and sea lion able to "solve for X" and ponder why? And of course, why isn't every human equipped to analyze brain slices for their long-term-potentiation potential? What's so great about not being brilliant?

Clearly mental simplicity is the preferred method. Most animals operate on 95 percent instinct and a smidgen of learning. Many operate on 99.99999 percent instinct, but have a few flexible neurons that can adapt to some important feature of the environment.

To appreciate the benefit of low intelligence in a typical animal, let's recruit an Einstein Mouse and a Lennie Mouse, set them loose in the backyard, and see how they fare.

Before they even build their nests, they raise the issue of fuel efficiency. Brain tissue is expensive. It's high in fat, which is high in calories. And operating a brain burns a lot of calories, as well. The human brain consumes one out of every five calories you burn in a day. So an animal with a big brain is forced to spend more time foraging, and that exposes him to risks. It's not clear whether a quick brain burns much more fuel than a slow brain, but that is certainly possible.

Now let the mice run. Their first task is to find shelter, perhaps in a hollow tree, or an abandoned bumblebee tunnel in the ground. Einstein may end up with the best shelter, because his curiosity will lead him to explore. He may consider the hollow tree, but then continue to investigate a bumblebee tunnel, then proceed to a crack in the stone foundation of my house. Einstein takes possession of weatherproof accommodations in a batt of R-13 fiberglass.

Lennie, meanwhile, found a bumblebee hole, and began lining it with dry grass. His house is nearly finished by the time Einstein sets to work modifying my insulation.

Results of Round One: Lennie got the job done quicker, which reduced his exposure to predators. He can move ahead to Round Two, storing food. But his shelter is vulnerable to flooding when it rains, and if he should entice a mate to move in, their babies will be vulnerable to the pawing of skunks. Einstein's shelter took longer to find, but it may last his lifetime. There is plenty of room to expand if he accumulates a harem, and skunks will not be a problem for his babies. But he's running behind schedule.

As Round Two opens, Lennie wastes no time finding a meal. He's satisfied with the basics. A few seeds from the flower bed will do. A bit of rotting fruit under the apple tree provides both water and a sweet treat.

He eats, stores a little extra, and he's ready for the next challenge, finding a mate.

Einstein emerges from the foundation and, like Lennie, gathers an easy mouthful of seeds from the flower gardens. But those extra dopamine receptors in his hippocampus may be tickling him a bit. Something pushes him to look farther, check elsewhere, learn something. So he branches out from his fiberglass home, trotting along beams and probing the walls. In my hallway his curiosity pays off: a bag of birdseed.

And so on. Lennie devotes his full attention to life's fundamental tasks. Because he explores only enough to get the job done, he avoids unnecessary risks and saves time. He will take the same approach to finding a mate. Lennie will never have the most fabulous shelter or the most fat-and-feisty mate. But he will perform the job of being a mouse with efficiency. And that efficiency will translate into a reliable rate of success.

Einstein may hit it big. His appetite for stimulation may lead him to great things. But curiosity does not discriminate. Just as quickly as he'll investigate a new flower in the garden, he'll approach a trap I've left for him, smeared with something that smells new and marvelous.

In humans, too, smart is as smart does. The human race has a smattering of absentminded professors who can solve for X but can't find their way home. And it has a smattering of people whose intellects require so little exercise that they are content going through the same few behaviors, day after day. And most of us fall somewhere in between, able to enjoy both a mindless walk in the woods, and the old riddle: If a tree falls in the woods and no one hears it, does it make a sound?

Openness Facet:

IMAGINATION

IMAGINATION INDICATORS	RARELY	SOMETIMES	OFTEN
I LOVE TO DAYDREAM	❏	❏	❏
I INVENT NEW THINGS IN MY MIND	❏	❏	❏
OTHER PEOPLE'S IDEAS INSPIRE ME	❏	❏	❏

THIS GIVES YOU A QUICK look at where you land on this facet. If your answers tend toward the "often" side, you're higher in that facet.

People with a strong imagination are able to stimulate their minds from within. By conjuring up stories and situations, they create their own waking dreams. They are interested in mental exploration and possibility. Someone with a less active imagination is satisfied with the reality at hand. She is at home with facts and has no need to complicate the situation with what-ifs.

Imaginative Mouse

Mice are hard pressed to demonstrate every facet of Openness. They do dream. Usually Mitzi and Maxi seem motionless when they sleep, except for the rise and fall of their ribs. But just the other day I heard a shuffling of the bank statements and peeked in to find Mitzi eating. Conked out and sprawled across the shreds, she lay still except for her jaws, which shuffled and snipped on their own.

So perhaps mice do imagine, running fantasies through their little brains. But for now, we can only speculate on the imaginative life of mice.

And that said, I will speculate on how a specific type of imagination in simpler animals may have laid the foundation for the complex worlds that humans can create. In fact, I believe I can make an argument that the animal instinct for self-preservation gives rise to beliefs in animals that equate to religion.

If you have no idea what I'm talking about, come to my kitchen and push a spoon off the counter. The house will shake as a seventy-pound dog scrambles away from the satanic silverware. All animals have an instinctive reaction to sudden noises. Deep in their DNA it is written that a sudden noise often means that a predator has crept very close and is now *leaping*.

The dog doesn't analyze such an event. A chair tips over or a spoon falls, and the noise is sufficient to convince his brain: That chair is animate, and it has intentions. There is a table in this house that he treats with great respect, based on its previous behavior toward him. He believes that table has particularly unpleasant goals. He imagines life and intentions where there are none.

In this sense, the dog is deeply religious. Mice, I expect, are equally devout.

Imaginative Human

Scientists find imagination interesting only when a person's inner world gets uppity and rejects reality. If your brain is that cut off from the outside world, your life is going to be difficult. But if you're just puttering along believing in fairies and UFOs, you'll be fine.

Magical thinking used to be considered a bad sign in a human. Researchers saw it as an inability to tell reality from fantasy. That sounded dangerously like psychosis. But these days more scientists view magical thinking as a normal side effect of having a brain. It's a universal feature of childhood. And most people hold on to irrational beliefs right through adulthood.

I'm not one of them. A few years ago a friend began quizzing people on their magical beliefs. Her quiz began with the most popular beliefs and worked down to the oddities:

Do you believe in a god?

Do you believe in heaven?

Do you believe in reincarnation?

Do you believe in ghosts?

Do you believe in UFOs?

Do you believe in the horoscope?

Do you believe in fate?

And so on. I scored zero. My personality is 100 percent magic free.

It was not always the case. I firmly believed in the tooth fairy, although I harbored sufficient doubts that I would seal my tooth in an envelope, then make tiny marks across the flap that would be disturbed if it were opened by means other than magic. My poor parents. And Santa Claus was the most marvelous entity in my little world. I fretted over the time-and-space problem presented by his obligation to visit every house on the planet at midnight. It seemed only marginally more feasible when my parents explained the concept of time zones. Nonetheless, I believed. When the doubt really kicked in, Dad carved a potato stamp and created reindeer footprints that trotted in a window, down the stairs, around the living room, and then back up and out. (In white paint. It's a feature of human development that kids don't grasp the difference between various materials—white paint, white snow—until a certain age.) But by the next year, the jig was up.

I was a normal kid. But I did not have a magically inclined personality. And my environment reinforced my genes. Neither of my parents held any magical beliefs, and as science-minded people they both preached the virtues of experiment and logic. I remember taking home the little red Bibles that some churches still distribute in public schools. Determined to recapture the magic of my youth through this grown-up Santa Claus, I would read, I would attempt prayer, but...nothing. It just didn't stick.

What was I missing? For one thing, my left lateral temporal region is probably on the slow side. A team in the UK has found that when they temporarily disable this area above the ear in a magical thinker, his ability to see the invisible dries up.

The form of delusion this group studied was one that most of us have experienced. I experience it in the shower: Within the white noise of the rushing water I suddenly detect the ringing of a telephone. I pause and listen harder. There it is again! But it's not. My brain is merely playing with patterns. In randomness, it is seeking order. And in the wilderness of the white noise, it manages to sift out something familiar.

Capitalizing on this, a series of studies have found that about half of normal, healthy humans will hear the song "White Christmas" in white noise if you plant the suggestion that it might be there.

In this particular study, though, subjects sat before a computer screen that presented them with a series of dot patterns. About half the patterns in the series would contain a hidden image, researchers told the subjects, just press the button when you see an image. In truth, none of the patterns held hidden images. This team didn't care who "saw" nonexistent images. They just wanted to know if shutting down a person's left-lateral-temporal area made a difference. Did it ever.

Using a magnet to disrupt a patch of brain behind the temple, researchers found they could cut people's magical thinking in half. As a group, the twelve subjects "saw" only half as many illusory images as they had seen with their brains unrestrained.

Curiously, previous work had shown that teasing the other side of the brain with magnets could do the opposite—bring on hallucinated sights, smells, and even the sense that someone or something unseen is present. And work with psychotic people has indicated that the left side of their brain is likely to be hyperactive. So perhaps these temporal lobes are the temples of magic in a human personality.

In our Age of Reason, magical thinking can look silly or illogical. But it has a legitimate purpose. All animals are bombarded with data as they move through the world. Some of it doesn't matter—the color of sand, the feeling of wind. But some of it does, like the two eyes gazing at you, the rhythmic crunch of footsteps in dry leaves, the patch of ripples in a cove that mean fish are schooling.

Without an ability to detect patterns, an animal would drown in data. Everything would seem to have equal importance. So animals have evolved methods to detect meaningful patterns in a sea of information. And judging from the experiment above, at least part of that job is performed in a person's left temporal lobe.

A few people are too good at finding patterns. Their brains detect patterns everywhere. Where there are no eyes gazing, they see eyes gazing. Where no footsteps crunch, they hear footsteps. In the clouds they not only see faces, they also see frowning faces, leering down at *them*.

Many more brains are middling at it. They're on the lookout for patterns, and, whaddya know, they find them! *"White Christmas"? Yeah, that sounds like "White Christmas"!*

Some people's brains have no talent at all. Perhaps mine is one of those. Such brains simply don't connect the dots as readily as others. Or it may be I have a tremendous tolerance for ambiguity. Whereas some personalities find relief in closure, in rules, in answers, others just don't care.

You have to admit it, sometimes a ripple on the water is just the wind, not a sign of great fishing. Sometimes the crackling leaves portend a small bird, not an axe murderer. A brain that can tolerate ambiguity has the patience to wait and see. When people press me over my shocking lack of gods, they often ask if I can explain this, or that: *How did the universe come into existence? What created life?*

I don't know. But I also don't care. I do not crave a resolution to riddles. I don't need an answer to my questions. My brain is perfectly comfortable with not knowing. It is just fine with ambiguity.

The inverse is true of a personality with a low tolerance for ambiguity: Resolution is where it's at. This personality feels uneasy with unsolved riddles. They're discomfited when the dots don't connect. *How did the universe come into existence? What created life? AIIEEEE!*

All people, like animals, had better pay at least some attention to ambiguity. To hear a leaf crackle and disregard it is foolhardy. To turn away without determining whether those eyes really are watching you, that's risky, too. The world is full of signs, some warning of danger, others offering opportunities. All animals must give some thought to patterns.

As to how one interprets those signs, that's a matter of personal preference. Humans seem to be born ready to assign motives to everything around them, just as my dog does. Children don't skip a beat when you ask them to explain the behavior of the world around them: Everything is conscious, and everything has an agenda. One of my favorite examples of this came out in a test of kids' preference for a scientific or emotional reason for why rocks are pointy. No, rocks aren't pointy because they're made of pointy minerals, the kids laughed. Rocks are pointy so animals won't sit on them.

And into adulthood most people will still reach for a magical explanation when they encounter a puzzling pattern. To one person, the plight of a homeless schizophrenic is a sign that he has been disrespectful to a god. To another, it means the man has been taken over by devils. Then again, perhaps he beat a child in his previous life, and is paying for his crime in this life. Or someone cast the evil eye on him when he wasn't wearing a protective charm. Take your pick.

Humans imagine such explanations in order to calm our brains. As long as the interpretation doesn't clash with any other phenomena, your brain will be at ease.

Evolution of Imagination

Imagination has many excellent uses that would have promoted its evolution. In fact, it may be the most powerful tool in the human mental kit. Through imagination we can apply what we've learned through past experience to predict what might happen in the future. It helps us to understand the motivation of other people. It permits us to visualize how we might solve a problem. And it's the foundation for that ultrahuman feature, artistic expression.

But what good is that subset of imagination, magical thinking? Well, frankly, it's good enough for other animals, so why shouldn't it work for us? When my big dog hears a spoon hit the floor, his brain does not pause to consider various possibilities. It leaps straight to the worst-case scenario: That noise is made by an assassin, and that assassin is after him. He scrambles for safety.

Turns out, when he tiptoes back from the other room, it was just a spoon. But that's OK. Better safe than sorry. This dog plays it very, very safe. He assumes anything that moves has a life force like his own. And with a few exceptions, he assumes that those life forces are intent on killing him.

And you know what? It's working! He's still alive! Presuming the worst is a perfectly reasonable policy, for mouse and man. When the squirrels in my yard see a shadow slide across the ground, they presume it is a hawk. It could in fact be a harmless crow, seagull, or airplane. But they don't risk much by assuming it's a hawk and running for cover. When I toss out an unfamiliar food for the crows, they don't alight beside it and chow down. The first time I gave them chunks of stale chicken they landed ten feet away and approached with great caution. From one foot away they feinted at the chicken and leaped back, over and over. What if that meat was only pretending to be dead, and intended to leap up and kill them? They assumed the worst. It's a reasonable policy.

Humans add another level of complexity to our imaginings. We sophisticated, social animals have very social brains. Using our own brains, we can "try on" the feelings and motives of other people. And we know that others are studying us, too.

This awareness makes our imaginary lives so much richer! If a man throws a spear at you, he must be angry. Hence, when the bison all vanish, they must be angry, too. Or when a lake turns poisonous, it must want us to go away. If it hasn't rained in months, the entity in charge of corn must be punishing us. When child after child dies in its mother's arms, we must have done... what? Oh, how the ambiguity rankles! *It must be... it must be... what's the pattern here? Someone must be up there playing games with us! Should we give him something? Sing him a song? Chant to him?*

Until our imaginings prevent us from reasoning out a better answer, this kind of magical thinking can offer some protection from the seemingly random world. If the lake gets angry and poisons us whenever we camp too close to it, then we'll stop making it angry. If people who eat carnivorous animals are often skinny and weak, then those animals must be sending us a curse inside their meat. The rumble of thunder means the sky is mad and will soon try to roast us with fire bolts, so we should hide.

It doesn't matter why you change your behavior, as long as you reduce your risk. To the extent that magical thinking helps us to identify patterns, and protect ourselves from danger, it's a good tool.

Magical thinking does become a handicap when it butts heads with reality—and wins. Usually this causes only a minor inconvenience. The common religious taboo on eating pork was presumably a result of that omnivorous animal passing along *Trichinella* worms and other parasites to humans. Today's factory-farmed pigs are parasite-free. There is no longer a pattern in which the people who eat pork get sick or lose weight. But the religious dictum holds. In this case, abstaining from pig products isn't going to hurt anyone.

But the taboo against consuming blood presents a more serious conflict. Observed most famously by Jehovah's Witnesses, it extends to receiving blood transfusions at a hospital. This taboo may have originated when a group of people recognized a pattern: Those who handled or drank animal blood sometimes got sick. The magical explanation that endures to this day is that receiving blood in a hospital would offend a god. But in this case, the magical view is causing people to die who could be saved through the logical process that is science.

So, the benefits of magical thinking have limits. When a belief helps you to avoid danger or reduces conflict, that's helpful. But when science and education produce a new interpretation of the world and your brain can't accept the new paradigm, that's a problem.

That's when a tolerance for ambiguity comes in really handy. The ability to update a belief when new information arrives has its own merits. Such a brain sacrifices the efficiency of routines and rituals, but gains the potential to find new solutions to life's endless challenges.

Consider the ancient practice of sacrificing food, wine, animals, or people to the gods. Maybe about half the time, it appears to work and you get the rain or grapes or offspring you were hoping for. Even with a 50 percent failure rate, a lot of people will continue to bring the sacrificial cattle, year after year. For some brains, it's less stressful to assume the gods are angry than to accept that the behavior of rain and grapes are riddles without answers.

But if your brain tolerates—even enjoys—imagining alternative explanations, you might pay more attention to the times when sacrifice fails. Seeking alternative patterns, you may realize that rain waxes and wanes in an eleven-year cycle. Interesting. Perhaps you will plant your crops accordingly—dryland plants for dry years, temperate plants for wet years. And perhaps you'll stop killing your cows in a vain attempt to change the weather.

A tolerance for ambiguity grants some flexibility to a personality. And that flexibility allows a person to update her belief system when it gets too far out of synch with reality.

It can, of course, go too far. If you're so flexible that no belief sticks for long, you may spend a lot of time playing catch-up: studying your new weather pattern, finding the crops best suited to it, and learning to grow those plants. I see a version of this in friends who are always discovering themselves. Again and again they find their passion. Tossing aside their previous plan, they bore into the new "me." They take classes and buy equipment and read a dozen books and then ... some new pattern catches their eye. *I think I was meant to raise organic llamas!*

But even then, no harm, no foul. Belief systems become truly dangerous only when they're so rigid or flexible that they don't allow a person to protect life and family. When magical ideas tell a parent that her child is devil-possessed, or mustn't have a blood transfusion, evolution has a chance to act. When a facet of personality is so extreme that it interferes with successful reproduction, nature simply allows that experiment to fall by the wayside.

Openness Facet:

ARTISTIC INTERESTS

ARTISTIC-INTERESTS INDICATORS	RARELY	SOMETIMES	OFTEN
I LOVE THE SMELL OF THE OCEAN	❏	❏	❏
I'M CREATIVE	❏	❏	❏
BEING IN NATURE RESTORES ME	❏	❏	❏

THIS GIVES YOU A QUICK look at where you land on this facet. If your answers tend toward the "often" side, you're higher in that facet.

A person who scores high on artistic interests isn't necessarily an artist, but does crave immersion in natural beauty and the creative arts. Beauty and the expressions of human creativity energize and inspire this personality. A person with low artistic interests finds inspiration elsewhere. She may appreciate beauty intellectually, but doesn't have to seek it out to feel satisfied.

Artistic Mouse

I didn't think mice would have much to offer on the subject of artistic interests, until I met John Roder. Now I understand that the much-borrowed adage about jazz is as true for mice as for men: If it looks good, it is good.

I had already seen decent mouse housing, at the Lesch lab in Germany. In noise-buffered rooms, mice enjoyed social living unless they were aggressive males. Except during experiments, they had food and water available every second of their lives. The food was formulated to satisfy their love of gnawing. They had soft shavings underfoot, and a paper towel to shred for bedding and amusement.

This guided my homemaking efforts when I took custody of Mitzi and Maxi. I got two females, since they're social. On top of their shredded paper I gave them a napkin to shred. But I also got them a wheel. At the pet store, mice were always galloping on a wheel. And as a science writer, I'm keenly aware of the research about how exercise benefits the brain. Both humans and mice evolved to do a fair amount of galloping in a day. If modern humans are any indication, it would not be ideal for a mouse to give that up. So that was it, at the start: condo, paper shreds, napkin, wheel.

But from time to time I would read something new about mice: *They love to climb.* And I would add a rope swing. They *do* love to climb!

They're stimulated by looking for their food. Well, they didn't use the food dish anyway—Mitzi kept all the pellets beside the wheel so she could dine on the go. I took to scattering them among the shreds.

They love fresh food. Oh, yes they do! Within a month I had abandoned the pellets. The girls get birdseed, cracked whole nuts, dog kibble, spinach leaves, grapes, green beans, banana chips, cranberries, crackers, bread, corn chips, something new almost every day.

They love to chew. Too true. I gave them twigs, but they weren't interested. Then, after finding a raccoon skull with mouse tooth marks etched over its cranium, I remembered that wild mice meet their calcium requirements by gnawing animal bones and antlers. I gave them a small bone that they've worked on ever since.

They're curious. Mitzi loves exploring my (messy) desk, and used to "ask" to be airlifted from the condo by climbing the rope swing and nosing toward the skylight. I would boost her out and let her romp. She would slow down after thirty minutes (my daytime is her sleep time), and spend longer seconds in my sleeves or under my hands as I tried to type. I'd airlift her back. But I wasn't always sure I knew what she wanted. So one day I tucked a strip of cotton towel down through the skylight, and draped the other end down to my desk. In no time she discovered she could come and go as she pleased. (On the desk. White mice like Mitzi have been domesticated for more than five hundred generations—the human equivalent of about ten thousand years. I don't think she'd last long if she got onto the floor, into the wall, and had to earn her own living.)

So now I realize that mice can soak up a lot of experience and stimulation. Researchers are figuring this out, too. Just as physical exercise strengthens the brain, so does mental exercise.

Roder realized this long ago. Because he's interested in intelligence, he thinks about how a mouse's environment might support or undermine its genetic potential. Furthermore, mice are tough to train for these experiments, because they are not fundamentally brilliant, like rats. Coaxing every potential IQ point out of mice would be helpful.

But even adding a single diversion to mouse condos is unlikely to happen. If every researcher who uses mice stipulated how their animals were to be kept, the mousekeeping staff would be overwhelmed. And so it's a one-condo-fits-all situation.

"I'd like to put in those empty toilet rolls," Roder says. "But it probably won't happen. Even the smallest change has to please all the researchers." And it's a lot of toilet rolls. A lab may house tens of thousands of mice. Each condo must be cleaned once or twice a week, and every week or two the lid holding the water and food must be cleaned. The mousekeepers who support the Lesch team in Germany go beyond the standard when they add paper for the mice to shred. And each time they clean each condo, they must transfer a bit of that shredded bedding to the new cage, and add a fresh towel. It may not seem like much, but multiplied by a few thousand cages, it adds up.

This discussion isn't over. Scientists have shown that mouse brains develop abnormally in the standard condos. Exercise alone can make those brains more robust. And the effect of adding mental stimulation can be shocking. In one experiment, just rotating new toys and shelters into a condo each week was enough to bring mice out of genetically induced Alzheimer's and raise their function to the level of normal mice. In another, a wheel and toys prevented middle-aged mice from becoming forgetful as they aged. In a third, mouse babies born in a cage with some embellishments were larger and more likely to live.

So clearly, mice respond to an environment that stimulates their senses. Judging from their behavior, they like it. I'm not saying that a toilet-paper tube or a poker chip is a mouse's idea of art. Rather, I'm proposing that art is a human's idea of a wheel or a tube or a paper towel. It shakes up the brain and keeps it flexible, just as exercise conditions the body.

Artistic Human

Because people don't develop personality disorders related to our artsiness, this facet isn't subject to much research. The exception is a tenuous link between artists and schizophrenia. Perhaps the two run in the same families. For years scientists have gnawed on this bone, but the relationship still hovers in the realm of possibility, not established fact.

The research trend lately is to try to link mere "creativity" with mere "schizotypy." This amounts to a watered-down version of the artist-schizophrenic connection. Creativity is measured by a person's drawing or storytelling, or professional poetry-writing, painting, and the like. Schizotypy is measured with questions such as "Do you believe people can communicate telepathically?" and "Do you sometimes think objects or shadows are people?" and "Do you have trouble making conversation?" It's easier to find a link in this wider pool of people.

One of the most creative personality researchers in the business, Daniel Nettle of Newcastle University in England, did a typical correlation study: Interviewing professional artists and poets, he determined that they do have higher schizotypy than the average Joe. (They also have more sexual partners, which could explain why genes that produce schizophrenia remain in wide circulation.) So, there you have it: Artistic people are a little more schizophrenic than the rest of us.

But...other studies find otherwise. They find that creativity is most closely related to the Conscientiousness factor. Or they show that creativity relates to a "divergent thinking style," or that it simply cannot exist without a passion for intellectual engagement.

Clearly the facet, artistic interests, does measure an element of human personality. It's one I know very well. After days of sitting at a computer desk staring at words, I develop true cravings for art. "I need to look at beautiful objects," I tell my husband, and I bolt for the craft galleries downtown. The inlaid woods, the glass paperweights with coral reefs inside, the branching necklace with pearl "pussy willows," the velvet jewelry bags with blousy silk flowers, the comical clock made with bicycle parts, all tumble into my brain, landing in a pile of color, texture, glitter, humor. I feel a nourishing so great it's almost physical. Restored, I can resume my colorless labors. (Well, my labors aren't entirely colorless. My walls are yellow and orange, and I'm wearing a rhinestone bracelet.)

My husband enjoys the gallery, but he can live without it. He has other artistic interests. While driving the other day, he yelped, "Look at those wheels!" I scanned the scene in alarm, thinking the wheels might be ours, somehow separated from our car. No. Traveling at 35 miles an hour, he had noticed an unusually striking set of wheels on a parked BMW. The apples of his eyes hurtle all around him whenever he ventures onto the street. I can only imagine how distracted I'd be if paperweights and bracelets were the size of automobiles, and zooming around the landscape.

As for people with low artistic interests...I may not know any. Among my friends who completed the personality inventory, the couple who coincidentally tied each other for the lowest score has on their wall one of my favorite artworks of all time. They go to movies and live theater, and she, at least, wears unusual and beautiful jewelry.

Maybe birds of an aesthetic feather flock together, so even my low-scoring friends have strong artistic interests.

Evolution of Artistic Interests

Why do humans care about art at all? Why do we love color and texture and the act of creating?

One theory argues that our artistic preferences are based in our habitat preferences. Our species evolved on the plains of Africa, this argument goes. As bands of humans wandered around, they bore in mind a set of features they needed in a landscape. This is a subconscious version of what you do when you go house-hunting: We need three bedrooms, a garage, an electric range.

Only, their needs were different. They needed wide vistas. This would allow them the advantage of seeing prey and predators in the distance. Thus they would have time to plan appropriately.

They needed green landscapes. Greenery implies regular rainfall. That, in turn, implies a steady supply of plants and animals to eat.

Of course, they also liked the look of some animals fattening themselves in that landscape. It was like having a supermarket in the neighborhood.

A water body was also essential. A large lake would suffice, but flowing water was less likely to turn toxic from the waste of humans camped around it. Humans therefore preferred rivers, and we especially treasured waterfalls, whose natural aeration made the water even safer to drink.

A sprinkling of trees was crucial, both for shade and for safety. Humans aren't well armed, in terms of tooth and claw. One of our best options in times of trouble is to run and hide. A treeless landscape can strike the human eye as barren and hopeless.

And that, according to the hypothesis, summarizes why a modern human responds to art. We bear a biological admiration for signs of fertility.

And it's true that paintings of deserts with white bones in the sand are a rare and rather newfangled phenomenon. Puzzling and viewed with suspicion are canvases painted with colored blocks, or entirely black. For the bulk of art history, art has portrayed either lush, fattening landscapes, or fat, healthy prey, or fat, healthy humans. Even today, to walk through my favorite craft gallery is to survey an ecosystem that's bursting with flowers, happy fishes, fat clay rabbits, porcelain fruits, trees on canvas, trees as coatracks, and trees as ear jewels. No animal in the average art gallery dares to show a rib, and the trees would be well advised to look as though they're merely winterized, not dead.

What does this say about the diversity of art-love in human personality? Perhaps the beauty seekers have brains that run a continuous search for suitable habitat. That would help to make sense of a stray bit of data that's been bothering me. In those oodles of studies seeking a connection between schizophrenia and artists, one theme pops out time and again. Yes, the Openness factor is associated with a creative personality. But so is the Neuroticism factor. Over and over, the creative people also turn out to be a little bit more Neurotic than average.

If an interest in beauty grew out of our species' need for a safe habitat, well, who better to worry about safety than a Neurotic personality? That's what we anxious personalities live for—scanning for danger, and worrying, and planning for the future! That's why we're here!

If that's really the case, then a personality with low interest in art might also be low in Neuroticism. When migrating to a new habitat, this personality would be less inclined to fret about the potential for bad water and starvation. Without so much as a tree count or a shiver of dread about the slimy water hole, the low-art personality would be suited to plunge into a new habitat and take his chances.

Even if all this is true, it doesn't explain why so many humans produce art. Why is it so common for us to create songs, dances, sculptures, paintings, stories, and beautiful cars?

Brain scientists aren't terribly interested in our artistic interests, but anthropologists are absolutely fascinated with the origins of art. We'll probably never know why humans began to doodle. Was it to impress the opposite sex? Or was it a mere side effect of having a brain so powerful that it could turn inward and contemplate its own existence? Was it a way to silently communicate about nearby animals or enemies?

The first two possibilities have impassioned supporters. The "mating mind" camp contends that we make art because it shows off our general intelligence, our mental horsepower. And true enough, to create an original story, painting, or song requires coordination of many brain parts. You must first imagine your artwork. Then you must use your hands and eyes to produce it, refining as you go. And, if Daniel Nettle's research on poets and artists is any indication, a brain that can pull this off is attractive: More people want to mate with professional artists than with the average person. On the other hand, if it's such a great way to win mates, why are humans and bower birds the only species that evolved it? True, many birds do show off with rhythmic dancing, and some whales and dolphins develop unique songs, so perhaps other critters display their creativity in ways that we can't detect.

The second possibility is that artistic behavior is a side effect of the human condition. The complexity of our brain allows for connections and behaviors that are neither necessary nor useful for survival. But neither are they harmful. And so they remain, and are gradually recruited to support other goals. Making art can in fact help you attract a mate. It can also knit your group together to defeat an enemy. (I'm thinking of the Christian cross, and the "peace" symbol, and the song "We shall overcome," and war dances, and wedding dances.) And once art is legitimately useful to a group of people, it could help that group to survive while groups that can't exploit these new ways of communicating are less successful.

I'm interested in the third possibility because it seems related to what other animals do. When a chimpanzee wants a bite of banana from another chimp, she doesn't open her mouth. Why not? She wants that banana in her mouth. She also doesn't hoot and holler. When a chimpanzee wants food from another, she symbolizes her wish: She holds out her hand. My backyard crows do something similar. It's not always desirable to use noises when you want something, especially if you have vulnerable babies in the nest. So when a crow wants some grooming from a partner, she symbolizes her wish: She sidles near him and bows her head. A puppy, when denied something it desires, will gambol and yap in a dance of frustration.

These social animals use symbols to vent the pressure of their inner wants. It's not such a big step to acting out a more complicated message, or to scratching a stick-deer in the dirt to silently tell your companion what you've just sighted.

I'm not proposing that pantomime is dance. I'm proposing the opposite: that dance is a symbolic gesture. That music is an audible form of unspeakable yearning or happiness. And that drawing allows me to lay my mind on something that I can't quite get my hands on. As a kid, I wanted to be a beautiful lady, and I wanted horses. And that's pretty much all I drew. Most boys I knew drew sports cars, or shootout scenes. We were young animals overflowing with want. We were chimpanzees holding out our hands. "The arts," as we practiced them, were a way to converse with our brains about the yearning.

And some people have more to discuss with themselves than others.

LIBERALISM

LIBERALISM INDICATORS	RARELY	SOMETIMES	OFTEN
SOME PROBLEMS DON'T HAVE SOLUTIONS	❏	❏	❏
NO GROUP OF PEOPLE IS BETTER THAN ANOTHER	❏	❏	❏
I QUESTION AUTHORITY	❏	❏	❏

THIS GIVES YOU A QUICK look at where you land on this facet. If your answers tend toward the "often" side, you're higher in that facet.

Liberalism describes a personality's relationship to social structure. A person with high liberalism doesn't feel obliged to follow rules unless she can see the fairness and reason in them. Nor does she adopt the opinions of the majority unless she has tried them on and found they fit. A personality low in liberalism appreciates the stability provided by unwavering rules and traditions. This person doesn't care to explore gray areas, and finds security in the notion that something is either right or it's wrong. The two personality types map roughly onto the political ideologies we call conservative and liberal.

Liberal Mouse

I think we squeezed out all the social conformity and rebellion that we can get from a mouse in the Agreeableness chapter. You could probably catch a glimpse of liberalism in chimpanzees and other primates, and perhaps in supersocial animals like dolphins, wolves, and crows. Chimps, at least, do seem to vary in their political style. Some are brutal dictators while others build coalitions. But mice are so simple, and we are so new at studying them, that I doubt we can divide their social behavior more finely than we already have. For a full description of mouse Agreeableness, see altruism (page 123), cooperation (page 103), and morality (page 133).

Liberal Human

The standard manual of personality disorders does not recognize as abnormal that condition we call being a liberal. Nor is there an entry for being a conservative. Yet what causes a person to be one or the other is a question that consumes the world's democracies each time they prepare for an election.

Little brain research has been done on the subject. Since political orientation is not a disorder, expensive MRI time won't be spent on it. Nor will geneticists breed up lines of capitalist and Communist mice. Without a disorder to treat, there's no reason to investigate the genetics and chemistry of political leanings. Hence the research that does occur on this personality facet is often led by political scientists or psychologists, not brain scientists.

Conventional wisdom says that a liberal is soft-hearted, intellectual, and disheveled. A conservative is traditional, cool-hearted, and organized. A liberal braids her hair, while a conservative irons her undies. A liberal questions authority, and a conservative questions newfangled theories. Does research bear this out?

To a large extent, yes. People who call themselves liberals also tend to score high on tests of curiosity and creativity. They enjoy "effortful cognition," as the geeky phrase goes. Wrestling with ideas is not a chore, but a satisfying way to interact with the world. They're Extraverts of the mind, attracted to new information and experiences.

As a result, it's the "bleeding heart" liberals who can look at an axe murderer and say, "Well, his mother abandoned him, his father beat him, and the health-care system didn't catch his psychotic tendencies at an early age. Yes, he killed someone. But he himself was spiritually murdered. Interesting, isn't it?"

Conservatives, by contrast, tend to score high on the orderly facet of Conscientiousness. They are more comfortable following tradition and convention than wandering off to explore new territory. They do not care for ambiguity, preferring the stability that comes from clear rules and regulations.

As a result, the tough-hearted conservative looks at an axe murderer and says, "I don't care if you paint me five square miles of context. He broke the law."

This hearkens back to the ancient dichotomy of approach/avoidance behavior. In some ways, the conservative personality errs on the side of avoiding risk, while the liberal approaches risky situations in the interest of gathering more information. It's too early for the data to clump together in a clear pattern, but the trends are thought-provoking.

The conservative dislike for ambiguity is particularly suggestive. Ambiguity is what you feel when there is no perfect solution to a problem. Accepting it is the mental equivalent of dithering around in the open field, just begging to be attacked by a predator. In many ways, it's safer to commit to a course of action than to linger for days, debating the merits of every possibility. Conservatives, on average, are more comfortable committing to a response just to end the torment of indecision. I can often feel this impulse drumming its fingers as I twist and turn in the shifting winds of a decision: *Just pick one and get out of the storm!*

Conservatives also tend to rank high on something called "death anxiety." I had never heard of this, so I guess it's one anxiety disorder I escaped. And it's not really a disorder. Apparently the mere idea of death causes some people to feel uncertain and out of control—anxious. Some studies suggest that death anxiety reflects a fear that life itself has no meaning. For someone who doesn't enjoy ambiguity, that could be a pretty distressing possibility.

On the other hand, most studies find that conservatives rank lower than liberals on Neuroticism overall. In fact it's the Neuroticism in a liberal that leads to a bleeding heart. Being quick to spot the bad news—sad and angry faces in a crowd, for instance—the liberal is wide open to the sadness and pain in the world. She sees it, and on a biological basis she takes it personally. Your pain is her pain, your sadness is her sadness. People with high anxiety and a tolerance for ambiguity see the darkness in the world, and are able to spend some time contemplating it.

The Agreeableness factor also seems to divide liberals and conserva-

.tives. A handful of studies point to a difference in how each personality type views fairness. The liberal personality may have a built-in aversion to inequality, preferring that every member of his community have equal rights. The conservative personality has a higher tolerance for inequality, especially when equality would interfere with a clear chain of command. The conservative prefers to be part of a stable hierarchy, with someone in charge. Messy and inefficient consensus meetings, where everyone is the boss, drive them wild.

Does the brain itself look different in a liberal or conservative? That's a new subject for science, in part because these differences don't debilitate and disable millions of people the way schizophrenia and anxiety do. Political views encapsulate some of the most fundamental differences in human personality, so we take them to heart. But as passionately as I may dislike the opinions of my political rival, her views don't actually constitute a mental illness.

Nonetheless, a few researchers have scrounged together some money to investigate the inner workings of a liberal or conservative. Their results endorse the notion that conservatives are respectably high in Neuroticism, with brains that prefer safety to exploration.

One such study relied on good old sweaty palms to measure anxiety's relationship to politics. A sensitive amygdala produces high anxiety; and anxiety makes your palms sweat. Sweat makes your palms better conductors of electricity. So measuring palm conductivity is a cheap way to compare people's anxiety over any given stimulus. The stimulus in this case was photographs. Mixed among thirty-three photos were a person looking terrified with a big spider on his face; a bloody face with a dazed expression; and a flesh wound crawling with maggots. My palms started to sweat just reading about it.

The conservative palms sweated like gangbusters. Compared to liberal palms, they swam with sweat and pooled with perspiration. Those were anxious palms, produced by anxious amygdala. Liberal palms were calm, cool, and collected. (These are averages, of course. There was a healthy slice of overlap, where people who deplore prayer in school, the death penalty, warrantless searches, and biblical truth produced prodigious palmar perspiration as well.)

The researchers also tested the startle response, by measuring how forcefully a person blinks in response to a sudden noise. It's another measure of anxiety. Again the conservatives flew the anxious flag. Bam! Their eyes slapped shut with nearly three times the force of liberals, for each startling blast of white noise.

The researchers don't claim that a jumpy amygdala causes conservative politics. But they speculate that a brain sensitive to danger might be more inclined to support policies that are "protective." Support for gun ownership, the Iraq War, the Patriot Act, "obedience," "patriotism," and military spending was high in this group. So perhaps that vigilant organ the amygdala is a mascot region for protective politics.

A different team probed how the two personality types handle the demonic Go/No-Go task. This is the computer game I attempted at a Minnesota lab that studies how impulsivity and drug addiction are related. Go/No-Go is the test where three Xs appear on the screen and you don't click the button. But when two Xs and one O appear, you do click. Unless a split second later a box appears around the O. By then, of course, you've started to click. Can you call back the impulse? It's awful. This test should be banned under the Geneva Convention.

The scientists used a shower cap of electrodes to monitor the liberal and conservative brains as subjects took the test, hoping to spot interesting patterns. They were particularly curious about the anterior cingulate, a belt of cells wrapped around one of the brain's deeper onion layers, and a region known to wrestle with conflicting information. When you ask a person, "Look at the word 'green' but tell me the color it's printed in [red]," his anterior cingulate fires up to sort out the conflict.

Liberals have animated anterior cingulates. The current that shot through their brains was nearly twice as powerful as the current in a conservative brain. These brains were giving everything they had to the Go/No-Go challenge. They weren't any quicker on the draw—they just unleashed all their firepower when a conflict arose. These brains were also better able to halt the movement of their fingers when that infernal box appeared around the O.

The liberal brain, researchers ventured, should be quick to notice when its habitual behavior is a bad match for the current circumstance. The conservative brain, by contrast, produces a more stable response. Perhaps that response isn't finely tuned to every situation, but it is predictable.

This test dovetails with the research showing that a liberal personality is more tolerant of ambiguity. These brains do appear to devote more attention to life's conflicts and confusion. Conservative brains, meanwhile, aren't as likely to explore ambiguity. They notice it in good time, but they don't go bananas to evaluate it, and they don't change their habits to accommodate it.

Obviously, most people are neither extremely stable nor extremely changeable. Most of us fall somewhere between the extremes. The data from the anterior cingulate makes a tidy demonstration: People's rating of their own politics produced an unbroken line spread across the entire spectrum. Graphed alongside that line, the activity of their brains mirrors it beautifully: Liberal politics? Liberal anterior cingulate. Middle-of-the-road politics? Middle-of-the-road anterior cingulate.

It's a mathematical memo: Personality is a spectrum, and as individuals we spread ourselves all up and down its length.

Evolution of Liberalism

A hunter is out scouring the landscape for something to feed his family. He would prefer a deer, since it's a familiar package of meat that's easy to butcher and transport. If he finds fruit or starchy tubers along the way, so much the better.

Near the outer limit of his usual hunting ground he finally discovers a herd of deer, and wounds one. Tracking the animal's blood spots, he ends up in an unfamiliar valley. But he's a guy, and he doesn't need directions. He shoulders his meat.

Wending his way homeward, the hunter spots an unfamiliar glimmer of orange in a tree. Peering up through the leaves, he confirms that it's a bumper crop of fruit. It looks ripe based on the color. But its skin has a glossy sheen he's never seen before. He shivers, remembering his last bout of food poisoning. He continues on his way. *Some fruits can poison you. Stick with the things you know are good for your family.*

His brother went out that morning, too. He saw no deer, though he explored for hours. But heading home, he spied a glimmer of orange in a tree. What a peculiar day. A totally new fruit. He climbed into the canopy, picked one, and rested against the trunk. He took the tiniest nibble of the skin, and waited. Neither acridity nor numbness spread across his tongue. He took the tiniest nibble of the yellow flesh below, too tiny to taste, but enough to burn or numb if it was going to. He bit out a bigger morsel. It was tart, with clear juice. In his throat the pulp left a scratchy feeling but after ten minutes he still felt fine. He filled his bag and headed home, eating an entire fruit on the way. *Funny, you never know how a day will turn out.*

Around the fire that night there was hearty, familiar deer meat and there was a completely new fruit, surprisingly tangy on the tongue. Somebody had already eaten a whole one, and he still felt fine. So dig in, kids!

And some of the kids did dig in. Bored with meat, they gorged on this fruit that shone in the firelight. Their mothers rolled their eyes at one another, knowing that some young liberals might spend the next morning holding their bellies and groaning. But it wouldn't kill them, and it was good to see them expanding their diets.

And some of the old folks hung back. They sucked on the deer bones and waved away the orange fruit. *I didn't get this old by eating any old thing that hangs off a tree.*

THIS GIVES YOU A QUICK look at where you land on this facet. If your answers tend toward the "often" side, you're higher in that facet.

Your good friend spills red wine on your carpet but doesn't tell you about it. When you find out, quick: How do you feel? If your personality is high in emotionality, you will have the answer on the tip of your brain. You know how you're feeling, and you are able to describe your state of mind to others. If you feel a little stumped, or merely "mad," you probably rank lower on emotionality. It's laborious to rummage through your own head and try to name the feelings you find there. Generally, you don't find emotions very interesting.

Emotional Mouse

Self-awareness was once a hallmark of the human species. Only humans were thought capable of contemplating our selves as a concept. We assumed that other creatures had such simple brains that navel-gazing was simply not an option for them. They had no inner life, only an outer life.

Now we know that some apes, elephants, magpies (cousin to the clever crow and raven), and probably dolphins can conceive of a self to the extent that they recognize themselves in a mirror. This may not mean those animals can identify their emotions. They may simply experience them, the way a baby does, like weather systems moving through. But it's possible that self-awareness allows them to take note: *Gosh, I'm blue. I really prefer happiness.*

What the navel-gazing species have in common is that we're all social animals. We build and maintain complicated relationships. We must remember that so-and-so is a friend, and that such-and-such can be a jerk if he catches you alone. It takes a lot of brain to manage such a social life.

A mouse does not have these problems. Mice certainly recognize their friends and family members by smell. But the ties that bind them are pretty frail. When changing the condo of research mice, keepers will often transfer a pinch of bedding from the dirty condo to the fresh one. In that bedding mice detect their social order. If it's omitted, they may be socially blinded and have to fight about their dominance status all over again. It's as though they've never met.

My desk mice, Mitzi and Maxi, have known me for nearly a year, and although I know all their favorite foods, they have no idea what I like. In fact, they often mistake me for food when I reach into the condo to deliver snacks or fix the wheel after they cram a nut under it. The same fingers they cuddle under when they grow weary of desk exploration, they smell entering the condo and invariably bite. This isn't a territorial bite. It's a gentle, testing bite. My beloved mice daily mistake me for a blueberry.

So, seeking self-awareness, not to mention emotional self-awareness, in a mouse is probably a fool's errand.

Emotional Human

Emotional awareness is a real phenomenon in people, however. One of the best arguments for it is alexithymia, an inability to understand or describe one's own emotions. What a bizarre form of blindness!

This isn't the "emotional intelligence" that John Roder's wife has and he lacks. That's just a garden-variety sensitivity to the emotions of people around us. But Roder's son with schizophrenia probably has alexithymia. And Roder probably will, if he doesn't already.

Roder has flapped and sloshed through an entire day of talking, and shows no fatigue. He's one of those precious scientists that science writers love: They are so comfortable in their fallible human skins that they can relax. Instead of cowering behind a desk and treating each question like an incoming missile, these treasured souls perk up for a good question, and say unorthodox things like "I don't know." Roder is like that. He loves a good think. And yet he still hasn't made reference to the fact that when he turns toward me in his office chair he's likely to twirl past me and, feet slapping, claw back to an interactive quadrant.

I can't bear it. I chose my line of work because I never have to stick a microphone in the face of someone whose house is burning and ask, "So, how do you feel?" Insensitivity to the pain of others has never been my strong suit. So I made professional choices that minimized the necessity of sticking needles into people and asking if it hurt.

I take seventy-five deep breaths and I ask, as flatly as I can, considering the hyperactivity of my amygdala and the degree of my social anxiety: "So, what's *your* neurological story?"

Lurch-lurch, swing of the chair. "Huntington's."

Ah. How poetic. It's one of those brain diseases where you can watch your familiar old personality warp. Slowly. You could take notes, if you were so inclined. Mice wouldn't mind these diseases. With no landmarks to show them who they used to be, what would it matter? Humans mind.

Humans mind even as their son is peeling off the well-worn path of normalcy into the swamp of schizophrenia. One in four people with Huntington's try to kill themselves.

"My father had it," Roder continues. "It was hard watching him, as it progressed."

Is that alexithymia right there? My guess is that watching his father slur and spasm and go socially deaf was more than "hard." I don't know John Roder. Maybe such things are merely "hard" for him. But I think that for most people, to know and love somebody as he becomes somebody else is: Agonizing. Gutting. Hellish. Seemingly unending. Complicated with guilt. And more.

Huntington's chorea, named for the staggering "choreography" that patients perform, is a genetic disorder, producing a protein that slowly kills brain cells. Roder's father having it meant Roder had a 50 percent chance. And of course the fact that Roder has Huntington's means his children are another roll of the dice. In a photo portrait on the wall, his daughter embraces his son, all sepia smiles and formality. I search the faces for signs. All I can see is a minor slant to the boy's eyelids that can make a person look sad. That's probably just my overheated amygdala. But that kid has reason to be sad. On top of the genes for schizophrenia he has 50/50 odds of developing Huntington's. Two personality-altering diseases.

This isn't necessarily a ghastly coincidence. In some families Huntington's and schizophrenia seem to travel together. In some cases, the psychosis of schizophrenia arrives first, followed by Huntington's; in others, the order is reversed.

What the diseases share is an erosion of the social behaviors so central to human beings. It's as though these brains revert to a solitary lifestyle where meshing harmoniously with others isn't necessary.

The Huntington's case is creepy. Even people who have the defective gene but haven't yet developed the symptoms are somewhat blind to the emotions that play across the faces of other people. Once the body reveals through its chorea, or dance, that the brain is losing control, the emotional blindness is more severe. The negative emotions of others— anger, fear, disgust—are particularly impenetrable. A person who has gone emotionally deaf may offend you and have no way to use your facial feedback to adjust his behavior.

As the mutated protein accumulates, the neurons of the brain commit suicide. The brain shrinks. The frontal areas, those that allow us to temper our impulses and assemble complex thoughts, grow weak. Hunger or anxiety or frustration can spur aggression that shoots out unhampered by concern for how it might strike others. A person who once empathized and nurtured his family now rages and hits. His personality sheds its social skills.

Schizophrenia is more complex. Whereas Huntington's results from a single mutated gene, schizophrenia arises from dozens, or even hundreds, of genes, plus an environmental insult like a troubled pregnancy or difficult birth. Rather than a one-gene-wonder such as Huntington's, schizophrenia resembles the multiformity of snowflakes. Although each person's case may look similar at a glance, no two are identical.

That said, when a personality skitters off the edge of sanity and is formally labeled schizoid, family members often realize that the person's social coordination had begun to deteriorate a few years prior.

Schizophrenia, like depression, is a multiple-choice diagnosis. But one of the essential symptoms to qualify is "social dysfunction." And alexithymia is a frequent contributor to the problem. These personalities are either unwilling or unable to analyze their own emotions. They also struggle to see life from someone else's perspective, to stand in someone else's shoes. And not surprisingly, they also fail to empathize—to feel in their own brains the emotions of other people.

In a social animal, that's not a recipe for success. In order to work cooperatively, a person must be able to process others' emotions. If I sat down for dinner with your family and helped myself to the food on your plates, your faces and words would signal that I wasn't cooperating in an acceptable way. Being fluent in those signals, I could change my behavior. But if I were blind to the language of emotion, I would continue to break the rules even as your fear and disgust grew. I would have no way of knowing I was out of line.

That's a pretty nasty personality disease. You're bashing your social life to bits and you don't even know it. Add on a few multiple-choice options like hearing voices, believing your own eyeball is an enemy that must be killed, and a memory that's gone to pieces, and your social prospects are dim indeed. Scientists strain to determine if depression is an integral part of schizophrenic disease or a result of the social estrangement. Meanwhile, four people out of ten whose personality takes the detour we call schizophrenia feel so hopeless that they try to kill themselves. Humans are social animals. Without connections, we crumple.

This is a long-winded way to demonstrate that emotionality is a real dimension of human personality. If it were taken away, you would miss it.

Fortunately, most of us can navigate the basic floor plan of our own emotions. Most of us can rattle off a crude status report on a moment's notice. With a little training or time, the average person can engage those frontal lobes in a survey of the deeper brain regions, split the hairs of blunt emotion. Most of us can distinguish between disappointment, frustration, and defensiveness, and happiness, awe, and appreciation.

For some people this all comes very easily and naturally. For others, it's work. It's not immediately obvious why either extreme would be useful. But research into yet another peculiar disorder hints at one possibility. We may need to be in touch with our own emotions in order to understand the emotions of others. Without that internal reference, we would be less able to figure out how to cooperate with (or manipulate) the people around us.

In a normal brain like mine, watching you receive a painful shock causes a mirror effect: My brain acts as though it's feeling your pain. Whether I'm shocked or you're shocked, the same areas activate. But analgia, a rare congenital defect, leaves some people unable to feel their own pain. When you shock them they feel nothing. So what happens in their brains when they see someone else in pain? Can they relate?

Only sort of. When these people see images of someone getting hurt, their brains appear to go through the same experience, as a normal brain would. But that reaction is subconscious. Their conscious analysis of the injury is different. They tend to say that the victim's pain is mild, and not very upsetting. The net effect is as though they're hard of hearing. Without a fully functional organ, they miss part of the message.

Research on sociopathic people reveals something similar. Sociopaths don't show much emotion, nor do they seem to believe that emotions are important, regardless of who's experiencing them. Their subconscious brains, when penetrated by the vision of MRI, are just as disinterested. They're tone-deaf to the emotions that cross other people's faces. What this suggests is that the two functions are connected: My ability to read my own emotions is linked with my ability to read yours.

Again, the vast majority of us are fluent enough to get by. Perhaps the most emotionally conservative people don't enjoy conversing with the most open people, but there are enough of us that we can all find comrades who operate in our own comfort range. The babblers find fellow babblers, and the silent types gravitate to companions who can be counted on not to talk about their feelings.

It's the loss of emotional literacy right in the middle of a life that I find so disturbing. One day you're puttering along with a family and friends and children and a dog, all of whom rely on your ability to empathize. When they cry or whine, they can count on you to care. But then one day you don't.

John Roder, for all his choreography, continues to enjoy effortful cognition and the process of scientific research. He oversees a busy laboratory and his desk supports skyscrapers of journal articles and reports. His name appears on half a dozen such publications every year. He appears to be tireless.

"My neurologist keeps pointing out that retirement is an option. But I enjoy working," he says, swiveling back to the skyscrapers. "It's part of my therapy."

Having plans for the future, having a sense of purpose, and moving among other people, those are all therapeutic—not just for people with Huntington's chorea, but for all of us. It's just that for people with Huntington's, they also reduce the likelihood that you'll take your own life.

This human self-awareness is a double-edged sword. We make a big deal of our ability to analyze ourselves. We treat it like a membership card to a terribly exclusive club. Most animals aren't allowed in.

But maybe they wouldn't want in. Maybe Mitzi Mouse, who spends her whole life immersed in the moment and never compares today's emotions with yesterday's, has a better chance of achieving euthymia: good emotions. ✗ June 6

Evolution of Emotionality

Did we evolve our ability to name emotions from the inside or the outside first?

It is useful to read your own emotions, so that you can control them before you alienate people. An infant who acts on pure emotion is just as quick to hit you as smile at you. That kind of emotional directness wouldn't work as a basis for social cooperation. Cooperation requires compromise. Compromise requires that a person be able to curb her own selfish desires. If you can't control your emotions, you can't compromise.

As a brain matures, it gains control of the emotions. An adult brain has checks and balances that automatically restrain some of our riskier behaviors. You may feel aggressive toward someone who slides into a parking space you were waiting for, but even as your emotion boils toward action, your Neurotic amygdala is warning you to guard your safety, and your Agreeableness is reminding you that social harmony is worth some sacrifice. However, it's not clear that you need self-awareness to do that. Most of us don't have to put much conscious effort into managing our emotions. A baby will hand you her toy if you reach out for it.

I'm inclined to think humans acquired our emotional literacy from the outside, that it gained a foothold in our brains because it allowed us to exploit other people. Having evolved the capacity to read emotions around us, we then used that ability to decipher the storm fronts and warm spells that pass through our own brains.

The fact that a person with analgia can empathize with others' pain even though he can't feel his own is a hint. It suggests that understanding another person could be even more important than understanding yourself. After all, you already know what you've got up your sleeve. The mystery that so often confronts you is what other people have in mind. Armed with that insight, you can adjust your plan to maintain your position in the social swirl.

That's the road other animals have taken. A mouse is presumably not aware of his own emotions when he meets a strange mouse. But he instinctively registers the other mouse's emotions. A sideways stance he knows instantly means trouble. The rattle of a tail also telegraphs forthcoming belligerence. But a mouse who rears and leans away is asking for peace. Mice don't think about this. Their brains automatically process the emotional data, and use it to produce an appropriate plan.

And my human brain does the same thing, automatically sifting the environment for the emotions of others. If you stood to my left and made a frightened face, your expression would flow through my optic nerve without me even being aware of it. "Heads up! Something's wrong!" my amygdala would whistle, and I would have no idea why I felt uneasy. The human brain subconsciously registers all sorts of emotional body language. As we go about our errands, our brains are checking the expressions on faces all around us. We never even realize it's happening, even as it alters our behavior.

A recent experiment in generosity lifted the lid on this system of subconscious surveillance. People at computer stations were making decisions on how much money to donate to a common pot, on the assumption that the total donations would be doubled and divided evenly among the players, regardless of who might have skimped. It proceeded as usual, with people giving about half of what they had to work with, until Kismet came along. This is a robot with big blue eyes. When Kismet gazed out of a computer screen in a player's peripheral vision, that player suddenly became more generous. Totally unaware that her brain had registered watching eyes, the player was newly motivated to look like a good citizen.

Hmm, this sounds rather cynical. We evolved our regard for others' feelings only so that we can profit from our insight?

Well, even the most social creature on earth has to reproduce, if generous genes are to persist. And you can't reproduce if you can't control sufficient food, water, shelter, and protection. A parent who won't fight for the resources that her offspring need may seem like a real sweetheart to the rest of the community, but to her offspring, she's a death sentence. Even social animals must maintain a selfish streak.

This would all be moot if we were solitary creatures like tigers. Solitary animals read each other's emotions, but they don't even pretend to use this information to maintain peace and love within their species. A tiger can read aggression or courting intentions in the body language of another tiger, but she uses her insight only to further her own agenda. She doesn't cooperate. She doesn't compromise. She exploits.

It's a striking feature of the human animal that we will throw half of what we have into a pot for the common good. Human empathy does support an astonishing degree of care-taking and cooperation.

And it's not even voluntary, the caring and empathizing and emotion-sharing that we do. When I think about John Roder, his self-aware and degenerating brain, and his disordered son, and the other members of his family, I could just weep. Their pain is snagged in my own brain like a rusty fish hook. And why? It makes no difference to my life whether the Roder family is happy or sad. I've met only one of them, and we'll never meet again.

That's the social brain for you. It has no "off" switch. It evolved in an animal that clumped together in groups of a couple hundred. Evolution, itself being blind, didn't foresee that humans would one day know thousands of individuals, and see millions more on television. The obligation to feel the emotions of every starving Ethiopian child and every drowning Bangladeshi mother can be burdensome.

This overload phenomenon illustrates how a low level of emotional engagement could be helpful. If you're not easily distracted by the emotions of others, you can stick to your knitting, focusing on the practicalities of day-to-day living. You spend less time polishing your connections to others. You free up your attention for more tangible needs. While I sit mourning the fate of John Roder and his family, John Roder is probably making his lurching way to the vending machine for more shrimp alkaline, continuing his steady pursuit of a cure for failing intellects everywhere.

So You Think You Might Be Open to Experience

Of all the personality factors, a married couple is most likely to resemble each other in their degree of Openness. This is based on a fairly slim stack of research, but it certainly makes sense.

For one thing, any long-term relationship necessitates a certain amount of manipulation. The common methods to convince another human to do your bidding include reason, charm, whining, insults, and threats. We each tend to have a personal favorite. People with high Openness prefer reason. And apparently they prefer partners who prefer reason. It stands to reason. If one person in the relationship uses logic but the other employs a less logic-based method, chafing will result.

Furthermore, our social values shape the way we want to interact in our community. That covers such hot topics as whether we want our children in public or private schools, whether and where we go to church, whether we give our money to the National Rifle Association or the National Audubon Society, and so on. People go to the mat over these issues. And if a couple is mismatched on them, chafing will result.

So I can see why Openness is a good factor to match in your partner. A spouse's touch of anxiety or his impulsive nature can be cute. But if his social values and thinking style are quite different from yours, chafing will result.

To recap, the low-Openness personality is conservative. Not politically conservative, necessarily. Conservative in the sense of living with restraint. A person with low Openness isn't given to whimsical dresses and wacky socks. She's a creature who finds comfort in routine, and security in a set of rules that don't shift with the wind. She has no need for guided meditation in a sweat lodge, or a course in interpretive dance. Her inner life is an uneventful one. Furthermore, she would prefer it if you kept your own messy emotions under wraps. If you're looking for someone to lead a complicated campaign, you might want to find someone in sensible socks who won't be distracted by emotional dramas.

The really Open personality has a brain that hungers for exercise, whether through sensory input in the form of music and art, or a slow walk through the colorful chaos of Chinatown, or a debate over the possible origins of empathy in humans and mice. This personality is at ease in the gray areas of politics and emotions alike, welcoming more and more detail in order to better understand each unique issue or person. In his hunger for more mental fodder he'll consider attending your church, learning Vietnamese, and letting his daughter paint his toenails pink. If you're looking for someone to create a new school system that will adjust to the brain style of many different children, you might look for someone wearing hand-knitted socks who seems to know what you're thinking before you even say a word.

Can Openness change? Or is this one of the factors we're stuck with until death does us part? Well, because millions of people aren't suffering from Overly Open disorder, or Stick-in-the-Mud syndrome, this factor really doesn't get much research attention. People don't strive to become more lost in thought, or take classes in staying home and reading a book. So we really don't know how mutable the personality factor Openness is.

A few studies hint at the "nurture" half of the nature/nurture equation. One paper suggests that a culture with a narrow range of roles to play will produce people with lower Openness. But a culture in which a person's education sprawls across years, where one encounters many lifestyles, will produce personalities with higher average Openness.

I suppose that's what we see when we compare two cultures such as Iran and Australia. In one, a growing brain learns only a few ways to be a member of society. Perhaps none of those ways really suits his personality. But in the other, that same brain would encounter a dozen religions, a hundred art forms, and a thousand different conversations in the streets of any city. If his DNA is inclined toward Openness, it will find encouragement here.

The other thing science has learned about Openness is that it fades with age. And a shockingly young age, at that. By age thirty, according to one study, we start to get sloppy in our reading of each other's sad and angry expressions. By sixty-five the majority of us are measurably impaired in reading emotion on faces, in voices, and in body language. And it's not just the emotional component of Openness that withers. The entire factor contracts.

It's as though all of us, with time, shed and leave behind some of the most human parts of personality. Gradually we pull in our emotional ropes, and close the shutters on our social passions. We all go a little deaf to each other's emotions, and a little blind to the world's beauty.

As our interests turn homeward, and our energies turn inward, the men's eyebrows grow longer, the women get a few cheek whiskers, and we all become more like the mice.

CONCLUSION

There are two guys in my household who make me crazy. And the minute I leave the house, I run into dozens, hundreds of additional people who make me crazy. And globally, the people who make me crazy number in the billions.

What is wrong with these people? Why can't they be more like me?

As I familiarized myself with the human brain, locating the regions from which personality arises, a pattern also arose. Nearly all of the hot spots of personality cluster deep in the center of the brain. This is the limbic system. Tellingly, it's also known as the "lizard brain."

So, the brain regions that house a personality are very, very old.

A similar pattern emerged as I came to know the chemicals that help one nerve cell communicate with the next. Neurotransmitters such as serotonin and dopamine may be even older than the lizard brain. They regulate basic systems like digestion and circulation throughout the body of animals. As animals evolved more complex nervous systems, the neurotransmitters took on new tasks in the brain.

Again tellingly, the basic neurotransmitters have been around so long that they're pretty stable now. The serotonin system of a mouse is much like my own.

In these old, very old, ancient brain systems was not where I expected to find the colorful and dynamic human personality. I thought more of it would be in the cortex—the thick, outer layer of a brain. The human cortex is huge, compared to that of a typical mammal. The prefrontal part, the lobe that fills your forehead, is one of the humanest parts of a human brain. And, yes, that prefrontal cortex (PFC) does participate in our personality. It works like a conductor to harmonize all the competing urges and impulses generated by the older regions. But without those older regions, the PFC would be a conductor without an orchestra.

Now, I had known that "simpler" animals have personality. Anyone who's had more than one pet knows that. But cats and dogs are relatively new species like us. What came as a surprise was that ancient species such as hermit crabs have personality. They're not all the same. After a scare, some individuals come back out of their shell quicker than others. And alligators have personality. Some mothers guard their babies more aggressively than others, or wait more patiently for a drinking zebra to come into range.

In these animals, personality is synonymous with "survival strategy." A crab may be a born risk taker, or a born wait-and-seer. An alligator may have a short-term strategy, leaping at every zebra and catching few, or a long-term strategy of saving her energy for the perfect ambush.

The surprise wasn't how simple and biological the human personality is. It was how complex and diverse mouse personality is. As long as no single strategy proves more successful than the others, evolution keeps all the options in the gene pool.

So that's why human personality dwells down there in the limbic system: We evolved facing the same challenges as mice. And again, because no one strategy is fail-safe, we remain a species with diverse personality.

Diversity has magical powers. Try installing a curtain rod using a toolbox full of fifty identical hammers. Then try it again with a toolbox with fifty different tools in it. Of course that's not magic. But it is very powerful. Nature says so. Evolution has proven, in every animal scampering around on the planet today, that one personality per species is not sufficient. It takes more. It takes a range, a spectrum, a spread, a diversity. And when it comes to an animal with a brain as complex as our own, the diversity is so pronounced you can't miss it.

Even if you could, you wouldn't want to. Not really. People may drive me crazy. So many moments of my life would be easier if everyone were like me! But many more moments would become harder, or impossible. My survival strategy is not sufficient to handle every challenge. Nature says so. Evolution proves it ten times a day in my household alone.

ACKNOWLEDGMENTS

I feel great fondness and gratitude for Klaus-Peter Lesch and courageous researchers like him. Conducting animal research these days demands more than a commitment to ease human suffering. It also requires a willingness to risk the violence of antiresearch terrorists. And a special few go further still, welcoming journalists into their laboratories in an effort to help the public understand why animal research must be done. Without the many hours and precious access granted to me by Lesch, Thomas Wultsch, Andreas Fallgatter, Heike Wagner, and others in that lab, this wouldn't be half the book it is.

Likewise Marilyn Carroll graciously threw (kicked) open the doors to her lab despite many years of harassment from extremists. John Roder lined up a private seminar for me, presented by a team of brilliant young minds coming up through his outfit. Inga Neumann also gave me full license to roam the halls of her domain, importuning researchers and graduate students at will. And I still smile at the memory of Marc Caron's worried expression when I finally showed up—without a ketchup bomb, but still… My deepest gratitude to all, as well as to the clone-worthy Ramona Rodriguiz, and Garet Lahvis for reviewing the material. Any errors that may remain are entirely my own.

And thanks to so many more for making time. Science is a double-time job, and I know the time you gave me was precious: Kelvin Lim, Bill Wetzell, Oliver Bosch, Steve Duffy, Justin Anker, Tatiana Lipina, Bechara Saab, Greer Kirschenbaum, and Viviane Labrie all strove to help me grasp the fundamentals of neuropsychopharmachiatry. With varying degrees of success.

An additional army of researchers from around the world shared their work and answered questions selflessly. Heroes, all, working absurd hours to conquer depression, drug addiction, schizophrenia, and other heartbreaking malfunctions of the human head. I'm privileged to hang around the periphery of such idealistic and intelligent people.

Eleanor Holmes, my brilliant and beautiful and low-Neurotic cousin, has again reminded me that my dad wasn't the only person on earth able to spot an inconsistency—or a double space, or a maltreated German noun—at a hundred yards.

My gratitude also to the friends who submitted to the personality test so that I might exploit their peculiarities. I can say again that I'm privileged to hang with such idealistic and intelligent people.

A special *clink* to the Ladies Toast and Boast Society, including the lady who left us, rising like a little champagne bubble into eternity.

Another special *clink* to my agent, Michelle Tessler, my sister in conscientious and dutiful hard work. And a *clink* as well to my editor, Jill Schwartzman at Random House, yet another tireless drudge! I can say it a third time: I'm privileged to work with such idealistic and intelligent people.

A ;) to Bea Lurker, the best, if only, writing coach I've ever had. I'm pretty sure she exists, but the whole e-conversation was pretty surreal.

And of course, my appreciation for the home team is vast. I'm not talking just about Mitzi and Maxi the desk mice, although those adorable little souls did make me smile every single day I entered my office. It was the human family who understood that my closed door doesn't mean I don't love you. It just means I'm in here *trying to figure you people out!*

BIBLIOGRAPHY AND SELECTED REFERENCES

BIBLIOGRAPHY

Crowcroft, P. 1966. *Mice All Over*. London: G. T. Foulis & Co. *Note: This is a real study of real mice doing their real thing. Invaluable.*

SELECTED REFERENCES

For references not found here, please contact the author at: info @hannahholmes.net

1. NEUROTICISM

Anstey, M. L., et al. 2009. "Serotonin mediates behavioral gregarization underlying swarm formation in desert locusts." *Science* 323: 627–30.

Canli, T., et al. 2002. "Amygdala response to happy faces as a function of extraversion." *Science* 296 (5576): 2191.

Canli, T., et al. 2004. "Brain activation to emotional words in depressed *vs* healthy subjects." *NeuroReport* 15 (17): 2585–88.

Canli, T., et al. 2005. "Beyond affect: A role for genetic variation of the serotonin transporter in neural activation during a cognitive attention task." *Proceedings of the National Academy of Sciences* 102 (34): 12224–29.

Canli, T., et al. 2006. "Neural correlates of epigenesis." *Proceedings of the National Academy of Sciences* 103 (43): 16033–38.

Canli, T, et al. 2007. "Long story short: The serotonin transporter in emotion regulation and social cognition." *Nature Neuroscience* 10: 1103–9.

Canli, T. 2008. "Toward a neurogenetic theory of neuroticism." *Annals of the New York Academy of Sciences* 1129: 153–74.

Congden, E., et al. 2007. "Analysis of DRD4 and DAT polymorphisms and behavioral inhibition in healthy adults: Implications for impulsivity." *American Journal of Medical Genetics, Part B: Neuropsychiatric Genetics* 147B: 27–32.

Distel, M. A., et al. 2007. "Heritability of self-reported phobic fear." *Behavior Genetics* 38: 24–33.

Etkin, A., et al. 2004. "Individual differences in trait anxiety predict the response of the basolateral amygdala to unconsciously processed fearful faces." *Neuron* 44: 1043–55.

Gutknecht, L., et al. 2008. "Deficiency of brain 5-HT synthesis but serotonergic neuron formation in *Tph2* knockout mice." *Journal of Neural Transmission* 115: 1127–32.

Haas, B. W., et al. 2008. "Emotional memory function, personality structure and psychopathology: A neural systems approach to the identification of vulnerability markers." *Brain Research Reviews* 58 (1): 71–84.

Haas, B. W., et al. 2008. "Stop the sadness: Neuroticism is associated with sustained medial prefrontal cortex response to emotional facial expressions." *Neuroimage* 42 (1): 385–92.

Hamer, D. 2002. "Rethinking behavior genetics." *Science* 298: 71–72.

Hariri, A. R., et al. 2002. "Serotonin transporter genetic variation and the response of the human amygdala." *Science* 297 (5580): 319–21.

Hariri, A. R., et al. 2005. "A susceptibility gene for affective disorders and the response of the human amygdala." *Archives of General Psychiatry* 62: 146–52.

Harkness, K. L., et al. 2005. "Enhanced accuracy of mental state decoding in dysphoric college students." *Cognition and Emotion* 19 (7): 999–1025.

Larson, C. L., et al. 2006. "Fear is fast in phobic individuals: amygdala activation in response to fear-relevant stimuli." *Biological Psychiatry* 60: 410–17.

Lecrubier, Y. 2006. "Physical components of depression and psychomotor retardation." *Journal of Clinical Psychiatry* 67 (S6): 23–26.

Lee, W. E., et al. 2006. "The protective role of trait anxiety: A longitudinal cohort study." *Psychological Medicine* 36: 345–51.

Lemke, M. R., et al. 2000. "Spatiotemporal gait patterns during over ground locomotion in major depression compared with healthy controls." *Journal of Psychiatric Research* 34 (4–5): 277–83.

Lesch, K.-P., 2007. "Linking emotion to the social brain." *EMBO Reports* 8: S24–29.

Lesch, K.-P., 2001. "Genetic perspectives on the serotonin transporter." *Brain Research Bulletin* 56 (5): 487–94.

Omura, K., et al. 2005. "Amygdala gray matter concentration is associated with extraversion and neuroticism." *NeuroReport* 16 (17): 1905–8.

Shipley, B. A., et al. 2007. "Neuroticism, extraversion, and mortality in the UK health and lifestyle survey: A 21-year prospective cohort study." *Psychosomatic Medicine* 69: 923–31.

Strobel, A., et al. "Genetic variation of serotonin function and cognitive control." *Cog Neuroscience* 9 (12): 1923–31.

2. EXTRAVERSION

Cyr, M., et al. 2005. "Magnetic resonance imaging at microscopic resolution reveals subtle morphological changes in a mouse model of dopaminergic hyperfunction." *NeuroImage* 26: 83–90.

Chen, C., et al. 1999. "Population migration and the variation of dopamine D4 receptor (DRD4) allele frequencies around the globe." *Evolution and Human Behavior* 20 (5): 309–24.

Dreber, A., et al. 2009. "The 7R polymorphism in the dopamine receptor D4 gene (DRD4) is associated with financial risk taking in men." *Evolution and Human Behavior* 30: 85–92.

Dreisback, G., et al. 2005. "Dopamine and cognitive control: the influence of spontaneous eyeblink rate and dopamine gene polymorphisms on perseveration and distractibility." *Behavioral Neuroscience* 119 (2): 483–90.

Eisenberg, D. T. A., et al. 2008. "Dopamine receptor genetic polymorphisms and body composition in undernourished pastoralists: an exploration of nutrition indices among nomadic and recently settled Ariaal men of northern Kenya." *Evolutionary Biology* 8: 173–84.

Gillihan, S. J., et al. 2007. "Association between serotonin transporter genotype and extraversion." *Psychiatric Genetics* 17 (6): 351–54.

Hardin, M. G., et al. 2005. "Reward and punishment sensitivity in shy and non-shy adults: Relations between social and motivated behavior." *Personality and Individual Differences* 40: 699–711.

Lakatos, K., et al. 2000. "Dopamine D4 receptor (DRD4) gene polymorphism is associated with attachment disorganization in infants." *Molecular Psychiatry* 5: 633–37.

Rodriguiz, R. M., et al. 2004. "Aberrant responses in social interaction of dopamine transporter knockout mice." *Behavioral Brain Research* 148: 185–98.

Sahakian, B. 2008. "The innovative brain." *Nature* 456: 168–69.

Salahpour, A., et al. 2008. "Increased amphetamine-induced hyperactivity and reward in mice overexpressing the dopamine transporter." *Proceedings of the National Academy of Sciences* 105 (11): 4405–10.

Schnabel, J. 2009. "Rethinking rehab." *Nature* 458: 25–27.

Trihn, J. V., et al. 2003. "Differential pychostimulation-induced activation of neural circuits in dopamine transporter knockout and wild type mice." *Neuroscience* 118: 297–310.

Viggiano, D. 2008. "The hyperactive syndrome: Metanalysis of genetic alterations, pharmacological treatments and brain lesions which increase locomotor activity." *Behavioral Brain Research* 194: 1–14.

Whiteside, S. P., et al. 2001. "The Five Factor Model and impulsivity: Using a structural model of personality to understand impulsivity." *Personality and Individual Differences* 30: 669–89.

Zhuang, X., et al. 2001. "Hyperactivity and impaired response habituation in hyperdopaminergic mice." *Proceedings of the National Academy of Sciences* 98 (4): 1982–87.

3. AGREEABLENESS

Aragona, B. J., et al. 2004. "The prairie vole (Microtus ochrogaster): An animal model for behavioral neuroendocrine research on pair bonding." *ILAR Journal* 45 (1): 35–45.

Aslund, C., et al. 2009. "Impact of the interaction between the 5HTTLPR polymorphism and maltreatment on adolescent depression: A population-based study." *Behavioral Genetics* 39: 524–31.

Bachner-Melman, R., et al. 2006. "AVPR1a and SLC6A4 gene polymorphisms are associated with creative dance performance." *PLoS Genetics* 1 (3): e42.

Bakermans-Kranenburg, M. J., et al. 2008. "Oxytocin receptor (OXTR) and serotonin transporter (5-HTT) genes associated with observe parenting." *Social Cognitive and Effective Neuroscience* 3: 128–34.

Beiderbeck, D. I., et al. 2007. "Differences in intermale aggression are accompanied by opposite vasopressin release patterns within the septum in rats bred for low and high anxiety." *European Journal of Neuroscience* 26: 3597–605.

Bosch, O. J., et al. 2006. "Prenatal stress: Opposite effects on anxiety and hypothalamic expression of vasopressin and corticotropin-releasing hormone in rats selectively bred for high and low anxiety." *European Journal of Neuroscience* 23 (2): 541–51.

Bosch, O. J., et al. 2007. "Brain vasopressin is an important regulator of maternal behavior independent of dams' trait anxiety." *Proceedings of the National Academy of Sciences* 105 (44): 17139–44.

Bosch, O. J., et al. 2009. "The CRF system mediates increased passive stress-coping behavior following the loss of a bonded partner in a monogamous rodent." *Neuropsychopharmacology* 34 (6): 1406–15.

Caldwell, H. K., et al. 2007. "Oxytocin as a natural antipsychotic: A study using oxytocin knockout mice." *Molecular Psychiatry* 14 (2): 190–96.

Carter, C. S. 2006. "Sex differences in oxytocin and vasopressin: Implications for autism spectrum disorders." *Behavioral Brain Research* 176: 170–86.

Chen, Q., et al. 2009. "Empathy is moderated by genetic background in mice." *PLoS ONE* 4: e4387.

Crockett, M. J., et al. 2008. "Serotonin modulates behavioral reactions to unfairness." *Science* 320: 1739.

Domes, G., et al. 2007. "Oxytocin attenuates amygdala responses to emotional faces regardless of valence." *Biological Psychiatry* 62: 1187–90.

Domes, G., et al. 2007. "Oxytocin improves 'mind reading' in humans." *Biological Psychiatry* 61: 731–33.

Ebstein, R. P., et al. 2009. "Arginine vasopressin and oxytocin modulate human social behavior." *Annals of the New York Academy of Sciences* 1167: 87–102.

Fletcher, J. A., et al. 2008. "A simple and general explanation for the evolution of altruism." *Proceedings of the Royal Society B* 276: 13–19.

Hamlin, J. K., et al. 2007. "Social evaluation by preverbal infants." *Nature* 450: 557–59.

Heinrichs, M., et al. 2009. "Oxytocin, vasopressin, and human social behavior." *Frontiers in Neuroendocrinology* 30 (4): 548–57.

Israel, S., et al. 2009. "The oxytocin receptor (OXTR) contributes to prosocial fund allocations in the dictator game and the social value orientations task." *PLoS ONE* 4 (5): e5535.

Knafo, A., et al. 2007. "Individual differences in allocation of funds in the dictator game associated with length of the arginine vasopressin 1a receptor RS3 promoter region and correlation between RS3 length and hippocampal mRNA." *Genes, Brain and Behavior* 7: 266–75.

Knapska, E., et al. 2006. "Between-subject transfer of emotional information evokes specific pattern of amygdala activation." *Proceedings of the National Academy of Sciences* 103 (10): 3858–62.

Kosfeld, M., et al. 2005. "Oxytocin increases trust in humans." *Nature* 435 (7042): 673–76.

Miller, G. 2008. "The roots of morality." *Science* 320: 734–77.

Moll, J., et al. 2006. "Human fronto-mesolimbic networks guide decisions about charitable donation." *Proceedings of the National Academy of Sciences* 103 (42): 15623–28.

Neumann, I. D. 2008. "Brain oxytocin mediates beneficial consequences of close social interactions: from maternal love and sex." In Pfaff 2008, *Hormones and Social Behaviour* (Heidelberg: Springer-Verlag).

Panksepp, J. B., et al. 2007. "Affiliative behavior, ultrasonic communication and social reward are influenced by genetic variation in adolescent mice." *PLoS ONE* 2 (4): e351.

Shirtcliff, E. A., et al. 2009. "Neurobiology of empathy and callousness: Implications for the development of antisocial behavior." *Behavioral Sciences and the Law* 27 (2): 137–71.

Slattery, D. A., et al. 2008. "No stress please! Mechanisms of stress hyporesponsiveness of the maternal brain." *Journal of Physiology* 586: 377–85.

Thompson, R. R., et al. 2006. "Sex-specific influences of vasopressin on human social communication." *Proceedings of the National Academy of Sciences* 103 (20): 7889–94.

Veenema, A. H., et al. 2008. "Central vasopressin and oxytocin release: Regulation of complex social behaviors." *Progress in Brain Research* 170: 261–76.

Waldherr, M, et al. 2007. "Centrally released oxytocin mediates mating-induced anxiolysis in male rats." *Proceedings of the National Academy of Sciences* 104 (42): 16681–84.

Walum, H., et al. 2008. "Genetic variation in the vasopressin receptor 1a gene (AVPR1A) associates with pair-bonding behavior in humans." *Proceedings of the National Academy of Sciences* 105 (37): 14153–56.

Young, L. J., et al. 1999. "Increased affiliative response to vasopressin in mice expressing the V_{1a} receptor from a monogamous vole." *Nature* 400: 766–68.

4. CONSCIENTIOUSNESS

Ahmetoglu, G., et al. 2009. "The relationship between dimensions of love, personality, and relationship length." *Archives of Sexual Behavior:* online.

Anker, J. J., et al. 2009. "Impulsivity predicts the escalation of cocaine self-administration in rats." *Pharmacology Biochemistry and Behavior* 93 (3): 343–48.

Bozarth, M. A., et al. 1985. "Toxicity associated with long-term intravenous heroin and cocaine self-administration in the rat." *Journal of the American Medical Association* 254: 81–83.

Carroll, M. A., et al. 2008. "Selective breeding for differential saccharin intake as an animal model of drug abuse." *Behavioral Pharmacology* 19: 435–60.

Carroll, M. A., et al. 2009. "Modeling risk factors for smoking and other drug use in the preclinical laboratory." *Drug and Alcohol Dependence* 104 (S1): S70–78.

Carroll, M. A., et al. 2009. "Delay discounting as a predictor of drug abuse." In Madden, J. G., et al., eds. 2009. *Impulsivity: The Behavioral and Neurological Science of Discounting.* Washington, D.C.: American Psychological Association.

Dalley, J. W., et al. 2007. "Nucleus accumbens D2/3 receptors predict trait impulsivity and cocaine reinforcement." *Science* 315: 1267–70.

Knafo, A., et al. 2007. "Individual differences in allocation of funds in the dictator game associated with length of the arginine vasopressin 1a receptor RS3 promoter region and correlation between RS3 length and hippocampal mRNA." *Genes, Brain and Behavior* 7 (3): 266–75.

Stice, E., et al. 2008. "Relation between obesity and blunted striatal response to food is moderated by TaqIA A1 allele." *Science* 322: 449–52.

Underwood, M. D., et al. 2008. "Family history of alcoholism is associated with lower 5-HT2A receptor binding in the prefrontal cortex." *Alcoholism: Clinical and Experimental Research* 32 (4): 593–99.

5. OPENNESS

Acerbia, A., et al. 2009. "Cultural evolution and individual development of openness and conservatism." *Proceedings of the National Academy of Sciences:* online.

Alford, J. R., et al. 2005. "Are political orientations genetically transmitted?" *American Political Science Review* 99 (2): 153–67.

Amodio, D. M., et al. 2007. "Neurocognitive correlates of liberalism and conservatism." *Nature Neuroscience* 10: 1246–47.

Behrens, T. E., et al. 2008. "Associative learning of social value." *Nature* 456: 245–49.

Beitel, M., et al. 2004. "Psychological mindedness and cognitive style." *Journal of Clinical Psychology* 60 (6): 567–82.

Bell, V., et al. 2007. "Relative suppression of magical thinking: A transcranial magnetic stimulation study." *Cortex* 43: 551–57.

Borg, J., et al. 2003. "The serotonin system and spiritual experiences." *American Journal of Psychiatry* 160: 1965–69.

Botwin, M. D., et al. 1997. "Personality and mate preferences: Five factors in mate selection and marital satisfaction." *Journal of Personality* 65 (1): 107–36.

Boyer, P. 2008. "Religion: Bound to believe?" *Nature* 455: 1038–39.

Burke, K. A., et al. 2008. "The role of the orbitofrontal cortex in the pursuit of happiness and more specific rewards." *Nature* 454: 340–44.

Buss, D. M. 1992. "Manipulation in close relationships: Five personality factors in interactional context." *Journal of Personality* 60 (2): 477–99.

Carney, D. R., et al. 2008. "The secret lives of liberals and conservatives: Personality profiles, interactions styles, and the things they leave behind." *Political Psychology* 29 (6): 807–36.

Charlton, B. G. 2009. "Why are modern scientists so dull? How science selects for perseverance and sociability at the expense of intelligence and creativity." *Medical Hypothesis* 72 (3): 237–43.

Clapcote, S. J., et al. 2007. "Behavioral phenotypes of *Disc1* missense mutations in mice." *Neuron* 54: 1–16.

Correa, B. B., et al. 2006. "Association of Huntington's disease and schizophrenia-like psychosis in a Huntington's disease pedigree." *Clinical Practice and Epidemiology in Mental Health* 2 (1): online.

Danziger, N., et al. 2009. "Can we share a pain we never felt? Neural correlates of empathy in patients with congenital insensitivity to pain." *Neuron* 61 (2): 203–12.

Fleischhauer, M., et al. 2010. "Same or different? Clarifying the relationship of need for cognition to personality and intelligence." *Personality and Social Psychology Bulletin* 34 (1): 82–96.

Gerber, A., et al. 2009. "Personality traits and the dimensions of political ideology." *Social Science Research Network:* http://ssrn.com/abstract= 1412863.

Kapogiannis, D., et al. 2009. "Cognitive and neural foundations of religious belief." *Proceedings of the National Academy of Sciences* 106 (12): 4876–81.

Kossowska, M., et al. 2003. "The relationship between need for closure and conservative beliefs in Western and Eastern Europe." *Political Psychology* 24 (3): 501–18.

Kempermann, G., et al. 1997. "More hippocampal neurons in adult mice living in an enriched environment." *Nature* 386: 493–95.

Koten, J. W., et al. 2009. "Genetic contribution to variation in cognitive function: An fMRI study in twins." *Science* 323: 1737–40.

Lehrer, J. 2009. "Small, furry... and smart." *Nature* 46: 862–64.

Lovestone, S., et al. 1996. "Familial psychiatric presentation of Huntington's disease." *Journal of Medical Genetics* 33 (2): 128–31.

Mill, A., et al. 2009. "Age-related differences in emotion recognition ability: A cross-sectional study." *Emotion* 9 (5): 619–30.

Nettle, D., et al. 2006. "Schizotypy, creativity and mating success in humans." *Proceedings of the Royal Society—Biological Sciences* 273 (1586): 611–15.

Oxley, D. R., et al. 2008. "Political attitudes vary with physiological traits." *Science* 321: 1667–70.

Saab, B. J., et al. 2009. "NCS-1 in the dentate gyrus promotes exploration, synaptic plasticity, and rapid acquisition of spatial memory." *Neuron* 63: 643–56.

INDEX

ABOUT THE AUTHOR

HANNAH HOLMES is the author of the *The Well-Dressed Ape*, *Suburban Safari*, and *The Secret Life of Dust*. Her writing has appeared in *National Geographic*, *The New York Times Magazine*, *Los Angeles Times Magazine*, *Discover*, *Outside*, and many other publications. She was a frequent contributor on science and nature subjects for the Discovery Channel Online. She lives in Portland, Maine.

ABOUT THE TYPE

The text of this book was set in Janson, a misnamed typeface designed in about 1690 by Nicholas Kis, a Hungarian in Amsterdam. In 1919 the matrices became the property of the Stempel Foundry in Frankfurt. It is an old-style book face of excellent clarity and sharpness. Janson serifs are concave and splayed; the contrast between thick and thin strokes is marked.